Goethe's
"Exposure of Newton's Theory"
A Polemic on Newton's Theory of Light and Colour

Goethe's "Exposure of Newton's Theory"

A Polemic on Newton's Theory of Light and Colour

Translated by

Michael Duck and Michael Petry

With an Introduction by

Michael Duck

Imperial College Press

Published by

Imperial College Press
57 Shelton Street
Covent Garden
London WC2H 9HE

Distributed by

World Scientific Publishing Co. Pte. Ltd.
5 Toh Tuck Link, Singapore 596224
USA office: 27 Warren Street, Suite 401-402, Hackensack, NJ 07601
UK office: 57 Shelton Street, Covent Garden, London WC2H 9HE

Library of Congress Cataloging-in-Publication Data
Names: Goethe, Johann Wolfgang von, 1749–1832. | Duck, Michael John. | Petry, Michael.
Title: Goethe's "exposure of Newton's theory" : a polemic on Newton's theory of light and colour /
    Michael John Duck (retired research chemist, Univ. of Munich and AMR Tech), Michael Petry.
Other titles: Zur Farbenlehre. English | Exposure of Newton's theory
Description: New Jersey : Imperial College Press, 2016. |
    Originally published in German in 1810 as: Zur Farbenlehre.
Identifiers: LCCN 2015048814| ISBN 9781783268474 (hc : alk. paper) |
    ISBN 9781783265886 (sc : alk. paper)
Subjects: LCSH: Color. | Light. | Newton, Isaac, 1642-1727.
Classification: LCC QC495 .G48413 2016 | DDC 535.601--dc23
LC record available at http://lccn.loc.gov/2015048814

British Library Cataloguing-in-Publication Data
A catalogue record for this book is available from the British Library.

In-house Editors: Harini/Catharina Weijman

Typeset by Stallion Press
Email: enquiries@stallionpress.com

Printed in Singapore

Contents

Introduction, *by Michael Duck*   vii
Preface to the First Edition of *Zur Farbenlehre* (1810)   xxxi
Exposure of Newton's Theory   xxxix
Appendix   237

Introduction

Johann Wolfgang von Goethe was well acquainted with Isaac Newton's work on refraction and his theory of light and colours. Indeed, he painstakingly repeated the experiments in Newton's *Opticks* (1704), which clearly demonstrate that light is heterogeneous. Yet he never abandoned his belief that light is immutable and that colours result from the interaction of light and darkness. It is argued here that the origin of Goethe's refusal to accept Newton's theory was not psychological, as is commonly supposed, but theological. It stemmed from his pantheistic belief in the spiritual nature of light, which he believed rendered it unknowable and not open to analysis.

Goethe was a phenomenalist.[1] That is to say, he believed that "one should seek nothing behind phenomena, for they themselves are the theory".[2] So, when he casually picked up a prism for the first time in January 1790 what he saw through it corroborated this thinking:

Goethe's first prismatic observation

I was in a completely white room. I expected, as I held the prism to my eye, mindful of the Newtonian theory, that the whole of the white wall would

---

[1] T. Holtsmark, 'Goethe and the Phenomenon of Colour' in *The Anatomy of Knowledge*, edited by M. Grene (Amherst, 1969), 47–71.

[2] Goethes 'Maximen und Reflexionen', in *Schriften der Goethe-Gesellschaft*, 21 (Weimar, 1907), No. 575, 124.

*look multicoloured, for the light returning from thence to the eye was sup-
posed to be shattered into many coloured lights. How amazed I was to
observe that the white wall showing through the prism appeared as white
as ever, and that only where something dark was present did a more or less
distinct colour manifest itself, the framework of the window appearing the
most vividly coloured, whilst the light-grey sky outside exhibited not the
slightest trace of coloration. It required little thought to realise that a
boundary was required to elicit the colours and, as if by instinct, I immediately said aloud to myself that the Newtonian doctrine must be
erroneous.*[3]

In other words, colours, surely, are not contained within light,
but result from mixing light with darkness.[4]

Goethe followed up this observation by looking through his
prism at a rectangular strip of paper on a black base.[5] Again, what he
saw was coloration at the white/black interfaces: a pair of "warm"
boundary colours, red and yellow, along the top edge of the strip, and
a complementary pair of "cool" boundary colours, violet and blue,
along the bottom edge; the two sets of colours being separated by a
now narrowed white strip in the centre.

Figure 1[6] shows Goethe's representation of the development of
the solar spectrum when a beam of sunlight is refracted through a
prism — it being formed (as he sees it) by the gradual overlapping of
the two sets of boundary colours.

The subsidiary coloured arrays shown along the base of the figure
represent the images seen on a piece of white card when this refracted

---

[3] J.W. von Goethe, *Konfession des Verfasssers*, written in 1810 and reproduced in *J.W.
Goethe, Farbenlehre*, edited by G. Ott and H. Proskauer, 3 vols (Stuttgart, 1979), III, 246.
[4] The modern explanation is, of course, that the white wall remains white in appearance, not because of the absence of darkness, but because every part of the retina is
still receiving every part of the visible spectrum — even after refraction of the incom-
ing light by the prism.
[5] For his own description see *Beiträge zur Optik, Erstes Stücke* (1791) reproduced in
volume II of the above J.W. Goethe, *Farbenlehre*, 28.
[6] This figure is an adapted version of Goethe's Plate V (see p. 37).

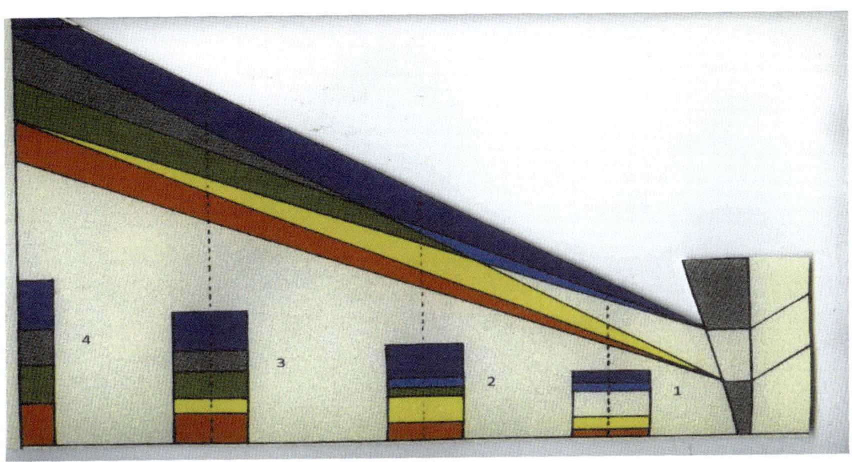

Figure 1   Goethe's objective spectrum.

beam is intercepted at intervals along its path. The first array in this sequence (No. 1) represents the initial appearance of the coloured fringes on the card and corresponds to the subjective observation[7] just described.

As Goethe retreated further from the strip, he noticed that the white centre gradually narrowed further until it finally disappeared, to be replaced by a now emerging green band (No. 2). Let us pause at this juncture to consider what the Newtonian assessment would be of this colour display: since dispersion of the differently refrangible rays by the prism has now been completed, we have achieved the fully resolved spectrum. Thus, even if one retreats further, its character will remain unchanged — except for its further elongation.[8]

---

[7] i.e., the "subjective" observation of a white card through a prism raised to the eye, as opposed to the "objective" observation (i.e., without looking through a prism) of the coloured image displayed on the card which is being illuminated by a prismatically refracted beam of light.

[8] See Figure 13 in the Appendix; and his explanation in pages 26–33 of Newton's, *Opticks*, 4th edition (London, 1730, reprinted New York, 1952). This is the edition that will henceforth be referred to.

Goethe, on the other hand, in accordance with his model, (illustrated in Figure 1), would expect, on retreating further, to observe a continuing diminution of the yellow and blue components, at the expense of green (No. 3).

It will no doubt come as something of a shock to Newtonians that this is precisely what does happen; as the observer retreats still further, the yellow band continues to shrink until it is scarcely discernible and the blue band disappears entirely (No. 4). The spectrum has effectively been reduced to Goethe's tri-colour. It would seem that Goethe has been vindicated, and *Newton* proved wrong – again! How can this be?

In fact, it so happens that this narrowing and eventual disappearance of blue and yellow from the prismatically observed white strip as the observer backs further away is due to a subtle physiological phenomenon:

One might reasonably assume that the observed hue of a coloured light source is independent of its brightness. But this is not necessarily the case. If, for example, an incandescent coil is seen through the dark red front glass of an infrared lamp, it appears to be more yellow in hue than its immediate surroundings — even though the spectral mix is the same across the entire image.

This change of hue with brightness was discovered independently by Wilhelm von Bezold in 1873 and Ernst Brücke in 1878 (N.B. long after Goethe's time) who gave qualitative descriptions of it[9] and it is therefore known as the Bezold–Brücke phenomenon. In 1931 Purdy[10] made quantitative measurements of it; he found that not only yellow but also blue light sources (but not green) undergo subtle shifts in hue to orange and mauve, respectively, if their intensity weakens. Thus, in the present experiment, as the observer retreats further, the yellow and blue bands, which are both affected by this phenomenon, are

---

[9]W. von Bezold, *Poggendorffs Ann, Phys. Chem.* 150, 22, (1873), 221–239. E.W. Brücke, *Sitzungsberichte der Akademie der Wissenschaften* (Wien) 77 iii, 221 (1878), 39–71.

[10]D. McL. Purdy, *Am. J. Psychol.*, 43, 541, (1931) and *Am. J. Psychol.* 49, 313, (1937). For more recent work see, for instance, G. van der Wildt and M. Bouman, 'The dependence of Bezold-Brücke hue shift on spatial intensity distribution' *Vis. Res.*, 8, 303, (1968) and R. Savoie, 'The Bezold-Brücke effect and visual non-linearity' *J. Opt. Soc. Am.*, 63, 10, 1253 (1973).

gradually absorbed by the green centre and narrow to the point at which they effectively vanish.

So, in short, when Goethe notes that, as the observer retreats, "The yellow and blue overlap increasingly until, eventually, the two colours have totally blended to form green, and the series reduces from red, yellow, green, blue, violet to red, green, violet,"[11] he is being factually accurate: his model happens to fit the described observations perfectly. Not for the reason he advocated; but he could hardly be blamed for that; he could hardly be blamed for not realising that it was sheer coincidence that the physiological Bezold–Brücke phenomenon, a phenomenon not consciously identified and characterised until long after his death, affects the appearance of the subjective spectrum in exactly the way that his totally unrelated physical theory predicts.

The above observations had a profound effect upon Goethe, and were to govern his thinking on light and colour for the rest of his life. They were to inspire him over the years to write dozens of poems on light and colour, and the Newtonian "heresy". Albrecht Schöne in his *Goethes Farbentheologie*[12] reproduces many of them. This one, written in 1827 is a typical example:

> *If you fitly can unite*
> *Bright and blackness, shade and light,*
> *Colour's secrets you may claim.*
> *Splitting light unique and holy*
> *Hold we for egregious folly*
> *Not to use a harsher name.*[13]

---

[11] *Outline of a Theory of Colours*, paragraph 216, see note 16.

[12] A. Schöne. *Goethes Farbentheologie* (Munich, 1987).

[13] *Einheit ewigen Lichts zu spalten*
Müssen wir für törig halten
Wenn euch Irrtum schon genügt
Hell und Dunkel
Licht und Schatten
Weiss man klüglich sie zu gatten
Ist das Farbenreich besiegt
The English translation is by E. Andrade, in 'Goethe as Natural Philosopher', *Nature*, 164 (1949), 338.

These initial observations were the springboard for Goethe to embark upon a massive investigation of colour phenomena of all kinds — and also to carry out a comprehensive study of the development of all colour theories from the time of Aristotle up to his own day. Furthermore, he proceeded to meticulously repeat many of Newton's experiments in order to re-interpret them according to his own theoretical approach.

Finally, a decade or so later he felt ready to publish his great work, in three volumes, which summed up all these endeavours, namely *A Theory of Colours* (1810).[14]

A Theory of Colours
Johann von Goethe (1810)

Preface

It is reproduced in full here,[15] because it presents us with a striking example of Goethe at his most idiosyncratic as he describes the layout of the book. For instance, it includes a satirical section where he likens Newton's theory, and the consequence of his refutation of it, to a derelict castle, and its eventual collapse.

Volume one: *Outline of a Theory of Colour*[16]

This so-called didactic part is an exposition of Goethe's own theory, as applied to all colour phenomena.

It is divided into six sections. The first three present the known colour phenomena (as Goethe defines them) under the three principle headings of physiological (with "relation to the eye" — after images,

---

[14] Goethe, Johann von: *Zur Farbenlehre*, (3 vols), Tubingen, (1810).

[15] Immediately after this Introduction. This 1840 English translation is by C.H. Eastlake. See note 16.

[16] *Entwurf Einer Farbenlehre*. This is available in a near contemporary (1840) English translation by Charles Lock Eastlake, which is reproduced in *Theory of Colours* M.I.T. Press, Cambridge, Mass. (1970). It will henceforth be referred to as *Outline*.

coloured shadows, pathological colours), physical ("with relation to light and darkness" — none of which exist without moderation and a boundary, and of which prismatic colours are but a sub-division), and chemical ("which we observe on bodies"). The subsequent three sections provide an overview of general notions derived from the particular phenomena presented in the first three sections; and an examination of the psychological and aesthetic effects of colour.

Volume two: *Exposure of Newton's Theory*[17]

The so-called polemic part.

Volume three: *Material for a History of Colour Theory*[18]

This is a study of the history of colour theory from antiquity to Goethe's day, and includes an extensive discussion and criticism of Newton and his theory. It is divided into seven chapters: The Greeks (mainly Aristotle), The Romans (Lucretius), The Interim (mainly Roger Bacon), The Sixteenth Century, The Seventeenth Century, The Eighteenth Century (first epoch) and The Eighteenth Century (second epoch). The sixth chapter concentrates upon Newton, whose theory is discussed in the sections entitled "Isaac Newton", *Lectiones Opticae*, "Letter to the Secretary of the London Society", "The *Opticks*", "Newton's relationship to the Society" and "Newton's personality".

There was no other history of chromatics in Goethe's day, apart from Joseph Priestley's *The History and Present State of Discoveries Relating to Vision, Light and Colours* (1772). In the *Preface* Goethe states that "With disgust and indignation we find Priestley, in his *History of Optics*, like many before and after him, dating the success of all researchers into the world of colours from the epoch of a decomposed ray of light, or what pretended to be so; looking down with a supercilious air on the ancient and less modern inquirers". But

---

[17] *Enthüllung der Theorie Newtons.*
[18] *Materialien zur Geschichte der Farbenlehre.*

that did not deter him from using G.S. Klügel's 1775 German translation of this work and taking copious extracts from it in the preparation of this third volume.

As might be expected, Goethe concentrated on the Ancients, the
Medievals and numerous seventeenth- and eighteenth-century theorists, and on their qualitative and philosophical statements about light
and colours. His interest, however, fell away as he approached the
closing decades of the eighteenth century, which saw physical optics
grow more exact and robust (i.e., more Newtonian).

If one reads *Outline* through Goethean eyes, that is to say, by
imagining light to be immutable, one finds that by-and-large his
descriptions, particularly when a prism is not involved, seem reasonable enough. Consider, for instance, his explanation of the colour
changes that take place at sunset. As the sun nears the horizon, its
light penetrates more and more of the atmosphere, which is "turbid"
(*Trübe*: Goethe's own word), leading to the sequence of ever-deepen-
ing colours: yellow, orange, red, and mauve.[19] Is this not a demonstration of colours being created by the adulteration of light with
darkness? Again, coloured paints are darker than white paint; is this
not because they contain a dark-related component?[20]
Thus, in essence, Goethe adheres throughout *Outline* to Aristotle's
view that colour arises from the transition of brightness to darkness.
Newton and his theory are never mentioned. By the time the reader has
followed all of Goethe's explanations of colour phenomena his thoughts
will have become quite divorced from the concept of wavelength.

If Goethe had stopped there and contented himself with simply
describing his own observations, one could understand — and generally accept — why he might be happy to ignore Newton's theory.
In the second, polemical part, however, he painstakingly repeats
the experiments of Book I of Newton's *Opticks*, vowing at the outset

---

[19] See *Outline* para. 154 for Goethe's own explanation.

[20] Goethe did not consider darkness to be merely an absence of light, but something positive. He tended to use the Greek word σχιερόν – probably so that it would stand out in
the text as representing something not to be ignored; see, for instance, *Outline* para. 69.

"not to let Newton get away with a single word, a single phrase".[21]
He was evidently confident that he would be able to re-interpret them
all in accordance with his own theory:

*"Since his assumptions are false and his whole exposition is erroneous,
his conclusion is also null and void. We hope to restore the honour of the
ancient philosophers, for although they did sometimes allow themselves to
be side-tracked, the broad outlines of the pre-Newtonian approach to the
phenomena were sound enough".* (366)

In *Opticks,* subtitled: *A Treatise of the Reflections, Refractions,
Inflections and Colours of Light,* Newton carried out 32 experiments, and
in *Exposure* Goethe repeats them all. About half are experiments carried
out with a single prism. These simple demonstrations in which light is
only refracted once are a great favourite with Goethean apologists,[22] for
they demonstrate the ability of light to create colour effects, without
unequivocally refuting the notion that it is immutable.[23]

However, whereas these earlier experiments were only performed
in order to demonstrate the heterogeneity of light, later ones were
devised with the specific objective of proving that:

*"The Phenomena of Colours in refracted or reflected Light are not caused
new Modifications of the Light variously impress'd according to the various Terminations of the Light and Shadow".*[24]

---

[21] *Exposure,* para 24. (Henceforth references to paragraphs in *Exposure* will simply be
given as numbers in brackets, i.e., here: (24)).

[22] For instance, N.M. Ribe's, 'Goethe's Critique of Newton: a Reconsideration', *Studies
in History and Philosophy of Science,* 6, 4, (1985), 315–335. Ribe's defence of Goethe
is based on his response to just two single-prism experiments from *Opticks* which deal
with (1) the boundary colours seen through a prism (Book I, Part I, Experiment 1)
and (2) a solar spectrum thrown by a prism (Book 1, Part I, Experiment 3); which, as
we have shown, can, in isolation, be plausibly explained by the Goethean approach.

[23] And, in his treatment of the Experimentum Crucis, where Newton employs two
prisms in creating a monochromatic beam (see Figure 18 in the Appendix), Goethe
does succeed in providing a plausible anti-Newtonian explanation based on colour-
blending effects (137).

[24] See Proposition I. Theorem I in Book I. Part II of *Opticks.*

That is to say, the experiments were expressly designed to disprove the Aristotelian notion upon which Goethe's own theory was based. The general consensus has always been that Newton succeeded brilliantly[25] and the scientific world was unanimous in its rejection of *A Theory of Colours* — particularly in Germany. Within a year of its publication, the German mathematician K.B. Mollweide in his book entitled *Presentation of the Optical Errors in Herr von Goethe's Theory of Colours and Refutation of his Objections to the Newtonian Theory* (1811) displayed a contemptuous mixture of sarcasm and condescension; whereas F.T. Poselgar declared that "great damage might be done to the science of optics unless Goethe's mistakes are thoroughly exposed".[26] And in 1813, C.H. Pfaff, in the most thorough criticism of *Zur Farbenlehre* ever published, declared that he felt impelled to expose Goethe's error "in the interest of science...since error, when it is taught by an otherwise excellent man, exercises its power over many minds, and even now important men are giving credence to the new theory".[27] The following year in England, Thomas Young wrote: "Our attention has been less directed to this work of Mr von Goethe by the hopes of acquiring from it anything like information, than by a curiosity to contemplate a striking example of the perversion of the human faculties".[28]

Yet Goethe remained unconvinced; he was never dissuaded from his conviction that light is immutable and that colour results from its

---

[25] For instance, in an address to the Royal Society in 1942, Lord Rayleigh asserted that "[Newton's optical experiments] are a model for all time of how experimental research should be conducted, and it is difficult for our generation to see how any intelligent person could refuse his assent". Lord Rayleigh, 'Newton as Experimenter', *Nature*, 150, (1942), 706–709.

[26] F.T. Poselger, 'Der Farbige Rand eines durch ein Biconvexs Glas Entstehended Bildes, Untersucht mit Bezug auf Herrn von Goethes Werk: *Zur Farbenlehre*', *Annalen der Physik*, 38, (1811), 151.

[27] C.H. Pfaff, 'Ueber Newtons Farbentheorie, Herrn von Goethes *Farbenlehre* und den Chemischen Gegensatz der Farben: Ein Versuch in der Experimentallen Optik', (Leipzig, 1813).

[28] T. Young, 'On the Doctrine of Colours by Johann Wolfgang von Goethe', *Quarterly Review*, 10/20, (1814), 427.

adulteration with darkness. Indeed, a mere four years before his death (Feb. 18th 1829) he told his friend J.P. Eckermann, that:

> *"As for what I have done as a poet I take no pride whatever in it...But that in my century I am the only person who knows the truth in the difficult science of colours — of that, I say, I am not a little proud, and here I have a consciousness of superiority to many."*[29]

How could this be? For us to attempt to answer this question it is instructive to turn to Goethe's treatment of some of Newton's key experiments.

Four experiments from Newton's Opticks

(1) *The demonstration that lights differing in colour also differ in refrangibility* (Book I, Part I, Experiment 2)

In this experiment, (one of the few where he uses no prism) Newton takes a card consisting of adjacent red and blue squares and covers it with thin black lines. It is then illuminated with a candle and the light reflected from it is passed through a convex lens, and focused on a screen which is moved backwards and forwards until the image is clear.[30] Newton finds that, when the lines on the red half are sharp, those on the blue half are blurred, and *vice versa*, thus demonstrating that the refrangibility of red light is different from that of blue light.

Goethe's response is to disagree with Newton's observations, and hence with the conclusion he draws from it:

> *"We have been unable to find any evidence of a position in which a less contrasting image is more distinct than a more contrasting one, and we are therefore obliged to categorise Newton's assertion as an appearance which is the result of wishful thinking and ingrained prejudice, and seen only with the mind's eye".* (76)

---

[29] J.P. Eckermann, *Conversations of Goethe with Eckermann,* translated by J. Oxenford (1850) and reprinted (London, 1930), 302.

[30] *Opticks*, p. 27, See Figure 12 in the Appendix.

Goethe maintains that the sharpness of the image does not depend upon the position of the screen, but only upon the contrast in brightness between the lines and the background. Basing his argument on this curious principle he flatly denies that, in any position of the screen, the lines on the blue part of the image (where the contrast is less) can be sharper than the lines on the brighter red part (where it is greater). One is reminded of Horatio Nelson's famous comment: "I see no ships".

(2) *The demonstration that sunlight consists of rays of different refrangibility* (Book I, Part I, Experiment 9)

In this experiment Newton demonstrates the heterogeneity of sunlight by using two prisms and employing the concept of the critical angle[31] (which is different for different colours). As the incident beam reaches the base of the first prism ABC at M some of it is reflected through the side AB to a second prism VXY where it is then refracted to tp. As the prism ABC is rotated anti-clockwise (Figure 2a), the

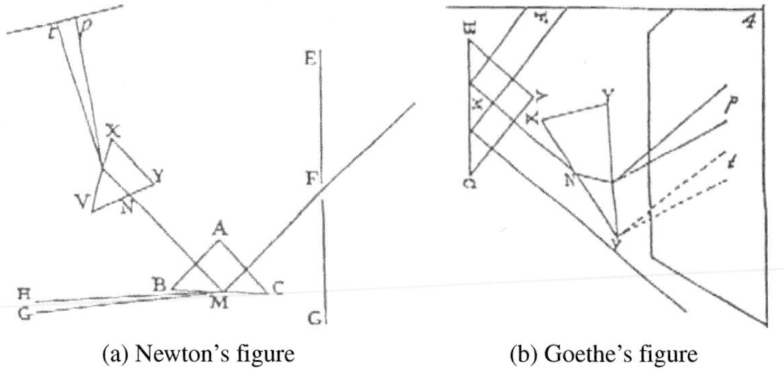

(a) Newton's figure   (b) Goethe's figure

Figure 2   The reflection of light at the critical angle (*Opticks*, Book I, Part 1, Experiment 9).

---

[31] When light enters a prism, as in Figure 2a, some is normally reflected and some transmitted. Since air is optically less dense than the material of the prism, however, the transmitted ray is bent away from the normal (to HG). Therefore, if the incident ray meets the rear face of the prism (BC) at less than a certain critical angle (i.e., here less than angle FMC), it cannot emerge at all, and is totally reflected internally towards the second prism XVY.

critical angle is reached first for the most refrangible rays MH causing violet to vanish from HG and the violet fringe of the image *tp* thrown by the second prism (at *p*) to become more intense. As the rotation continues, each colour in the spectrum HG disappears in sequence, whilst the intensity of each colour in *tp* increases in sequence. Finally, the refracted image HG disappears entirely. From this:

> *"'tis manifest, that the Beam of Light reflected by the Base of the Prism, being augmented first by the more refrangible Rays, and afterwards by the less refrangible ones, is compounded of Rays differently refrangible".*[32]

Goethe rejects this explanation. Providing a figure of his own, and ignoring the concept of the critical angle, he argues that as the prism ABC is gradually rotated clockwise (Figure 2b):[33]

> *"at first, only the upper edge of the image* pt *will display itself in blue and violet on the screen opposite; only somewhat later, when the whole image N has been intercepted by the prism VXY, will the lower edge t appear".* (200)

In other words, Goethe argues, the disappearance of the colours at *t* (in fact Newton only observed a change in intensity, not a disappearance of the colours at *t*) is due to the fact that only part of the beam must have been intercepted at that time. But, of course, intercepting only part of the beam could never result in the disappearance of the red/yellow fringe at *t* (even according to Goethe's theory, which predicts that, since the edge at V is a boundary, it must create boundary colours). It would only result in the image *pt* being narrower and fainter. Furthermore, Goethe's description takes no account of the events that occur at BC.

In short, Goethe's explanation makes no sense at all.

(3) *The demonstration that colours produced by refraction are not caused by boundary effects* (Book I, Part II, Experiment 1)

---

[32] *Opticks* p. 55.
[33] The figure is No. 4 in Goethe's Plate VIII.

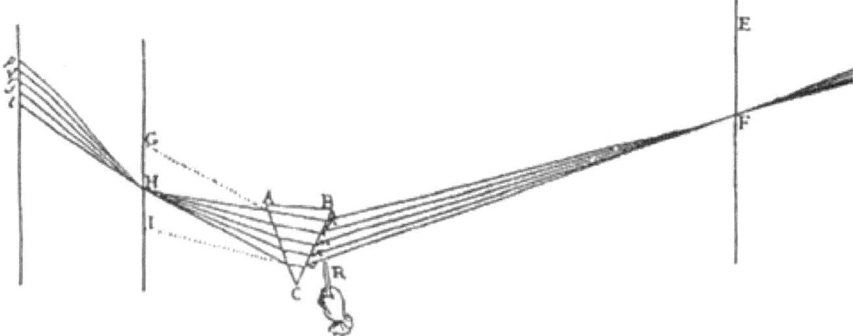

Figure 3   Newton's waving rod experiment (*Opticks*, Book I, Part II, Experiment 1).

A beam of sunlight from F is projected through the prism ABC. The still white central part of the beam is allowed to pass through hole H in a black screen on to a white sheet, where it paints the spectrum *pt* (Figure 3). If a rod is placed in the path of the incident beam, the spectrum loses some of its intensity, as might be expected, since some of the incident white light is being blotted out; and also, some of its coloration. But if the rod is now waved around across the beam the missing colours reappear, and others disappear in their place, i.e., different colours disappear for different positions of the rod. This demonstrates that, although boundaries (supplied by the rod) are present, they have nothing to do with the colour changes which occur; it is the screening out of certain rays (i.e., the subtraction of certain rays from the beam) which causes the colour changes. This phenomenon can only be explained by assuming that light is heterogeneous. However, Goethe would have none of it. He simply makes the trivial comment that the colour changes result from additional coloration provided by the coloured fringes thrown by the shadow of the rod.[34] However, the rod is *in front of* the prism; therefore the light playing on the leading face of the prism is colourless at all times. This demonstrates that it is not colour fringes from the rod that cause the colour changes, but the blotting out of some of the variously refrangible rays.

---

[34] See (360).

So, again, Goethe's explanation makes no sense — not even according to his own theory.

(4) *The demonstration that the colours of the spectrum can be mixed to re-create white* (Book I, Part II, Experiment 10)

In this experiment Newton disperses a beam of sunlight with a prism to create a coloured spectrum on a sheet of white paper. He then refocuses it with a lens, when the image becomes white again. This demonstrates that white can be re-created by mixing colours. Goethe's response is to avoid involvement:

> *"Since this experiment belongs to the heterogeneous group in which prisms and lenses are used in combination, it can only be covered properly in the Supplementary Section which we have often referred to".* (372)

Goethe eventually produced an essay in 1822 entitled *In Lieu of the Promised Supplementary Section.*[35] In it, however, he only deals with coloured lights, the visual and chemical effects of coloured lights on minerals and plants, and the action of coloured glasses; experiments employing lenses are not discussed.

However, Newton then carries out a variant of the experiment in which the reconstitution of white is effected with two prisms instead of a lens (Figure 4). Since here no lens is employed, Goethe can no longer use the above excuse to defer treatment of the issue. So he ignores it; it is not even mentioned by him, let alone analysed. Newton had provided this figure, so Goethe could hardly have overlooked it. It is not surprising that he did not wish to deal with this experiment, for its implications are fatal to his theory, according to which a second refraction, even if in a contrary orientation to the first one, must add to the adulteration of the beam, and therefore must result in an *increase* in coloration.

---

[35] 'Statt des Supplementaren Teils', written in 1822 and reproduced in J. W. von Goethe, 'Die Schrifen. zur Naturwissenschaft', *Deutsche Akademie der Naturforscher (Leopoldina)*, 2 divs, 15 vols, edited by R. Matthaei *et al.* (Weimar, 1947), 17, 21–39.

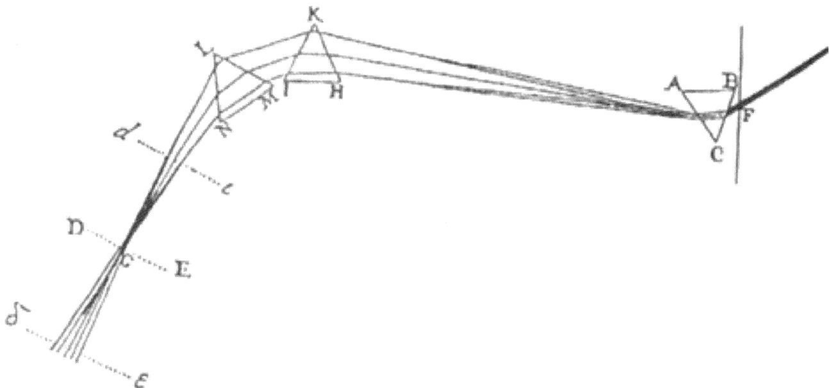

Figure 4   The reconstitution of white with prims (Opticks, Book I, Part II, Experiment 10).

One might think that Goethe would have finally realized by now that his theory was untenable; that Newton was right after all. Yet, he commences his *Conclusion* to *Exposure* with the opening statement that:

> *"We believe we have now fulfilled our task of providing a polemical treatment of the first book of the* Opticks, *that we have clearly demonstrated the extent to which Newton's hypothetical explanation and derivation of the coloured phenomena caused by refraction are untenable".*

What the foregoing clearly shows is that, in reality, Goethe had no intention of taking Newton's arguments seriously. If an experiment that cast doubt on his own theory could not be re-interpreted in a pro-Goethean light, a treatment of it was shelved, or it was simply ignored. *Exposure* purports to be a scientific document, but whenever the facts presented by Newton were unpalatable, the principles of scientific enquiry were suspended.

Given that so many of Newton's refraction experiments clearly demonstrate that light is not immutable – and that it is, in fact, composite and contains coloured lights, darkness playing no part in the creation of colour – how could Goethe, who repeated the experiments, maintain that they were a sham? He was not persuaded by them to abandon his own theory. Why was this? What was it that closed his mind to Newton's logic?

There have been many suggestions. Apart from the conjecture that Goethe simply did not understand the experiments,[36] various psychological explanations have been proposed; such as "a blind hatred of Newtonianism",[37] "self-deceit",[38] and even "paranoid psychosis"(!).[39] It is true that the mere thought of the existence of Newtonians angered him: he wished that "they wore special garments, so that they could be distinguished from sane people" (572); accused them of trickery (113) and lumped them all together as "Newton's flock" (654).

But, none of this answers the question (whichever description of Goethe's behaviour you choose from the above): what was the root cause of his antagonism? What was it that sustained Goethe in his battle against his legion of accusers? Why would he never abandon his belief that Newton's theory was "hocus pocus" (113)? If he refused to accept that light is composite, how did he visualise his immutable light as an entity?

Goethe's concept of light

Unfortunately, Goethe never passed any opinion on what he considered to be the physical nature of light. In fact, he opens his *Preface* to *A Theory of Colours* with the following statement:

"It may naturally be asked whether, in proposing to treat of colours, should not light itself first engage our attention?" To which we get the resounding answer: "It is useless to attempt to express the nature of a thing (i.e., light)

---

[36] "It seems strange that Goethe should have allowed himself to be wrought up into a state of mind which prevented his seeing clearly what could be seen by instructed school children". B. Chance in 'Goethe and the Theory of Colours', *Annals of Medical History*, 5, (1932), 357–365, (371).

[37] G.A. Wells, 'Goethe's Scientific Method and Aims in the Light of his studies in Physical Optics', *Publications of the English Goethe Society*, 38, (1968), 69–113.

[38] "Without reading the polemical part of the *Farbenlehre* one cannot get a real feeling of the psychological abyss of Goethe's self-deceit". S. Jaki, 'Goethe and the Physists', *American Journal of Physics*, 37/2 (1969), 195–203 (198).

[39] K.R. Eissler, *Goethe: Eine Psychoanalytische Studie*, 1775–1786, 2 vols (Basel, 1985), II, 1256.

abstractedly". Furthermore: "We should try in vain to describe a man's character (here, a metaphor for 'light'), but let his acts (the colour world) be collected, and an idea of his character (i.e., the behaviour of light, as opposed to what it actually consists of) will be presented to us". In other words, we can fully characterise the world of colour (by which light manifests itself) but, any attempt to unravel the mystery of light itself is futile.

Even when he discusses the opposition of the wave theorists Malebranche and Euler to Newton's corpuscular model, (that being an opportunity for him to join forces with two anti-Newtonians) he declines to take sides (457–458). However, when he does talk about light *per se*, the adjectives he employs do give us a reminder of how he thought light should be treated as a concept. For instance:

"Now that the author believes he has fixed firmly in our minds the idea that our pure, white, <u>simple</u> light".[40]

And:

"Light, this <u>essence</u>, which we are aware of only as a unity, as indivisible."[41]

Implicit in these descriptions is the notion that there is something wondrous, almost sacrosanct, about light.

Further:

"Now the author turns to the true source, namely the sun, as being a light we readily regard as being <u>Primordial</u>."[42]

---

[40] *Nachdem der Verfasser uns genugsam uberzeugt zu haben glaubt, dass unser weisses, reines, <u>einfaches</u> helles Licht.* (my emphasis) (194).

[41] *Das Licht, jenes <u>Wesen</u>, das wir nur als eine Einheit, als einfach wirkend gewahr werden* (my emphasis) (27). *Wesen* is used when reference is made to the inner, particularly spiritual, nature of a thing.

[42] *So wendet sich der Verfasser an die rechte Quelle, zur Sonne namlich, als demjenigen Lichte, das wir gern fur ein <u>Urlicht</u> annehmen* (my emphasis) (82).

This last quotation is reminiscent of the views of the fluid theorists of the time who believed, after *Genesis*, that light emanates from God and that, on the Fourth day, He created the sun to be the means whereby light would carry His message as it expanded throughout the Universe. Thus, such "Light metaphysicians" saw light as having spiritual properties. For instance, the English theologian John Hutchinson (1674–1737) propounded a system for the Universe that was based upon the notion of a "subtle" fluid, which appeared in three guises: **fire, light**, and **air** (or spirit). The sun was the great source and reservoir of this fluid, where it was in the condition of **fire**, from whence it perpetually emanated to the outmost bounds of the Universe in the condition of **light**; and from thence it was returned condensed in the condition of **air** or spirit, being melted down at the sun into fire once more. The perpetual circulation of this fluid accounted for all the phenomena of the heavens. This system, which was expounded in Hutchinson's *Moses's Principia* of 1748, employed no mathematics, on the grounds that "the Heavens cannot be measured by man" (p. 222). Like Goethe, he was strongly anti-Newtonian whom he considered to be a man "living in a box, peeping out at a window, or letting light in at a hole: or in separating or extracting the spirit from light, which can scarce happen in Nature".[43]

The notion that Goethe, too, saw light as something spiritual, is more explicitly suggested by various poems and aphorisms he wrote, such as:

"There is nothing new in the [misguided] *notion that the separation of God's most unique and pure creation* [light] *into parts has been achieved".*[44]

---

[43] See: *The Philosophical and Theological Works of the Late Truly Learned John Hutchinson Esq.*, edited by R. Spearman and J. Bate. 12 vols (London, 1748); A.T. Kuhn, 'Glory or Gravity: Hutchinson vs. Newton', *Journal of the History of Ideas*, 22/3 (1961), 303–322.

[44] *Neu ist der Einfall doch nicht, man hat ja selber den hochstcn Einzigsten reinsten Begriff Gottes in Teile geteilt.* Written in 1795 and reproduced in A. Schöne. *Goethes Farbentheologie* (Munich, 1987), 181.

Goethe's theology

Throughout his life Goethe expressed deep religious convictions, although he never associated himself with any of the traditional denominations. Fundamental to his faith, however, was a pantheistic belief in God's omnipotence throughout Nature. Most eighteenth-century theologians made a clear distinction between God and His creation. Thus, all power was seen as residing in God, whereas Nature was seen as functioning in accordance with the forces that God had set in motion within it the so-called Clockwork Universe. Goethe, on the other hand, saw God and Nature forming a unified polarity of being.

"With me, a pure, deep, innate, and constantly followed conception has been the view that God is inseparably within Nature, and Nature inseparably within God; and this idea has been at the basis of my whole existence."[45]

i.e., since Nature is part of the unity that is God "to perceive her in all her beauty was to Goethe the most supreme of all human privileges – to perceive and marvel over, but not to manipulate, to reduce to a mathematical formula – to disturb its unity".[46] Certainly, Newton's attempt to isolate light and split it into parts was to Goethe quite unacceptable:

Triumph of the School

"What an inspired idea! The immortal master [i.e., Newton] *demonstrates to us how to split up the ray that we have always known as being simple".*

---

[45] Entered by Goethe in his diary in 1816 and reproduced in Sherrington, *Goethe on Nature and on Science* (Cambridge, 1942), 30.
[46] *Ibid* 26.

Scepticism of the observer

"That is a pious hope! Just as the Church has for long divided God into three, so he divides light up into seven."[47]

Goethe's various utterances about Christianity cannot be rendered consistent even by the greatest scholarly ingenuity; but it seems here that, as a result of the Pietist phase in his life (1768–1771), he became highly critical of established and institutional Christianity[48] and, as an Arianist,[49] he correlated "the heresy of the Trinity" with "the heresy of splitting immutable light into seven parts". In short, God created the sun to illuminate and power the world, so that man could live in it and marvel at His works. However, the mechanism that allowed him to do this, i.e., light, must remain part of the mystery of God himself.

Conclusion

Goethe never abandoned his attempt to resuscitate the pre-Newtonian belief that light is immutable. The reason for this has never been satisfactorily explained. It is true that Goethe's theory can be used to give a plausible explanation of some of the more basic colour phenomena Newton studied – such as the boundary colours observed through a

---

[47] *Triumph der Schule*

Welch erhabner Gedanke! Uns lehrt der unsterbliche Meister, Künstlich zu teilen den Strahl, den wir nur einfach gekannt.

Zweifel des Beobachters

Das ist ein pfaffischer Einfall! Denn lange spaltet die Kirche Ihren Gott sich in drei wie ihr in sieben das Licht.
Written in 1795 and published in *Schillers Musenalmanach* in 1797. See, also, Schöne, 180.
[48] H. Loewen. 'Goethe's Response to Protestantism'. *Canadian Studies in German Language and Literature*, vii (Bern. 1972), 155.
[49] It would appear from the above and other quotes that he was sympathetic to the Arianist doctrine that Christ was not divine.

prism, or the spectrum formed when a beam of sunlight is passed through a prism. And, as we have seen, he used not a little ingenuity in some of them. However, most of Newton's more sophisticated experiments, all of which were repeated by Goethe, clearly demonstrate that light is composite and contains within it variously refrangible coloured lights. Goethe vigorously opposed this concept until the end of his life; notwithstanding the fact that he could find not a single physicist to support him. That he was able to continue to do this even after he had absorbed the contents of Newton's *Opticks* had nothing to do with physics. It stemmed from his unshakable pantheistic belief that light is inseparably bound up with God and, as such, is unknowable, not open to analysis — in either sense of the word.

Michael Duck
London, 2016

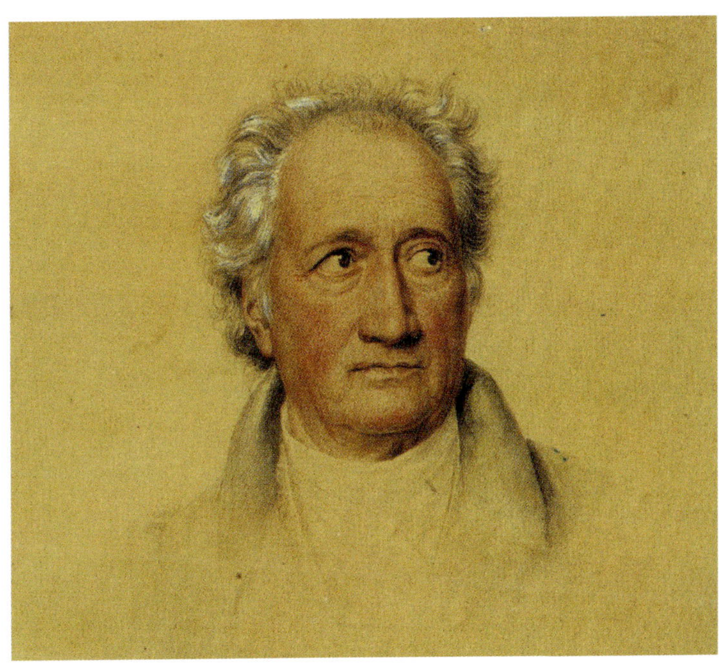

Johann Wolfgang von Goethe (aged 79)
A 1828 portrait by Josef Karl Stieler
Reproduced by kind permission of the Trustees of the Staatliche Graphische Sammlung München

Preface to the First Edition
of *Zur Farbenlehre* (1810)

It may naturally be asked whether, in proposing to treat of colours, light itself should not first engage our attention: to this we briefly and frankly answer that since so much has already been said on the subject of light, it can hardly be desirable to multiply repetitions by again going over the same ground.

Indeed, strictly speaking, it is useless to attempt to express the nature of a thing abstractedly. Effects we can perceive, and a complete history of those effects would, in fact, sufficiently define the nature of the thing itself. We should try in vain to describe a man's character, but let his acts be collected and an idea of the character will be presented to us.

Colours are acts of light; its active and passive modifications: thus considered we may expect from them some explanation respecting light itself. Colours and light, it is true, stand in the most intimate relation to each other, but we should think of both as belonging to Nature as a whole, for it is Nature as a whole which manifests itself by their means in an especial manner to the sense of sight.

The completeness of Nature displays itself to another sense in a similar way. Let the eye be closed, let the sense of hearing be excited, and from the lightest breath to the wildest din, from the simplest sound to the highest harmony, from the most vehement and impassioned cry to the gentlest word of reason, still it is Nature that speaks and manifests her presence, her power, her pervading life, and the vastness of her relations; so that a blind man to whom the infinite visible is denied, can still comprehend an infinite vitality by means of another organ.

And thus we descend the scale of being, Nature speaks to the other senses — to known, misunderstood, and unknown senses: so speaks she with herself and to us in a thousand modes. To the attentive observer she is nowhere dead nor silent; she has even a secret agent in inflexible matter, in a metal, the smallest portions of which tell us what is passing in the entire mass. However manifold, complicated, and unintelligible this language may often seem to us, yet its elements remain ever the same. With light poise and counterpoise, Nature oscillates within her prescribed limits, yet thus arises all the varieties and conditions of the phenomena which are presented to us in space and time.

Infinitely various are the means by which we become acquainted with these general movements and tendencies: now as a simple repulsion and attraction, now as an upsparkling and vanishing light, as undulation in the air, as commotion in matter, as oxidation and deoxidation; but always, uniting or separating, the great purpose is found to excite and promote existence in some form or other.

The observers of Nature finding, however, that this poise and counterpoise are respectively unequal in effect, have endeavoured to represent such a relation in terms. They have everywhere remarked and spoken of a greater and lesser principle, an action and resistance, a doing and suffering, an advancing and retiring, a violent and moderating power; and thus a symbolical language has arisen, which, from its close analogy, may be employed as equivalent to a direct and appropriate terminology.

To apply these designations, this language of Nature to the subject we have undertaken; to enrich and amplify this language by means of the theory of colours and the variety of their phenomena, and thus facilitate the communication of higher theoretical views, was the principal aim of the present treatise.

The work itself is divided into three parts. The first [didactic part] contains the outline of a theory of colours. In this, the innumerable cases which present themselves to the observer are collected under certain leading phenomena, according to an arrangement which will be explained in the Introduction; and here it may be remarked, that although we have adhered throughout to experiment, and throughout considered it as our basis, yet the theoretical views which led to the arrangement alluded to, could not but be stated. It is sometimes unreasonably required by persons who do not even themselves attend to such a condition, that experimental information should be submitted without any connecting theory to the reader or scholar, who is himself to form his conclusions as he may list. Surely the mere inspection of a subject can profit us but little. Every act of seeing leads to consideration, consideration to reflection, reflection to combination, and thus it may be said that in every attentive look on Nature we already theorise. But in order to guard against the possible abuse of this abstract view, in order that the practical deductions we look to should be really useful, we should theorise without forgetting that we are so doing, we should theorise with mental self-possession, and, to use a bold word, with irony.

In the second part [the polemic part] we examine the Newtonian theory; a theory which by its ascendancy and consideration has hitherto impeded a free inquiry into the phenomena of colours. We combat the hypothesis, for although it is no longer valid, it still retains a traditional authority in the world. Its real relations to its subject will require to be plainly pointed out; the old errors must be cleared away, if the theory of colours is not still to remain in the rear of so many other better investigated departments of natural science. Since, however, this second part of our work may appear somewhat dry as regards its matter, and perhaps too vehement and excited in its manner, we may here be permitted to introduce a sort of allegory in a lighter style, as a prelude to that graver portion, and as some excuse for the earnestness alluded to.

We compare the Newtonian theory of colour to an old castle, which was at first constructed by its architect with youthful precipitation; it was, however, gradually enlarged and equipped by him according to the exigencies of time and circumstances, and moreover was still further fortified and secured in consequence of feuds and hostile demonstrations.

The same system was pursued by his successors and heirs; their increased wants within, the harassing vigilance of their opponents without, and various accidents compelled them in some places to build near, in others in connection with the fabric, and thus to extend the original plan.

It became necessary to connect all these incongruous parts and additions by the strangest galleries, halls, and passages. All damages, whether inflected by the hand of the enemy or the power of time, were quickly made good. As occasion required, they deepened the moats, raised the walls, and took care there should be no lack of towers, battlements, and embrasures. This care and these exertions gave rise to a prejudice in favour of the great importance of the fortress, and still upheld that prejudice, although the arts of building and fortification were by this time very much advanced, and people had learnt to construct much better dwellings and defences in other cases. But the old castle was chiefly held in honour because it had never been taken, because it had repulsed so many assaults, had baffled so many hostile operations, and had always preserved its virgin renown. The renown, this influence lasts even now: it occurs to no one that the old castle has become uninhabitable. Its great duration, its costly construction, are still constantly spoken of. Pilgrims wend their way to it; hasty sketches of it are shown in all schools, and it is thus recommended to the reverence of susceptible youth. Meanwhile, the building itself is already abandoned; its only inmates are a few invalids, who in simple seriousness imagine that they are prepared for war.

Thus there is no question here respecting a tedious siege or a doubtful war; so far from it we find this Eighth Wonder of the World already nodding to its fall as a deserted piece of antiquity, and begin at once, without further ceremony, to dismantle it from gable and roof downwards; that the sun may at last shine into the old nest of rats and owls, and exhibit to the eye of the wondering traveller the labyrinthine, incongruous style of building, with its scanty, make-shift contrivances, the result of accident and emergency, its intentional artifice and clumsy repairs. Such an inspection will, however, only be possible when wall after wall, arch after arch, is demolished, the rubbish being at once cleared away as well as it can be.

To effect this, and to level the site where it is possible to do so, to arrange the materials thus acquired, so that they can be hereafter again employed for a new building, is the arduous duty we have undertaken in this second part. Should we succeed by a cheerful application of all possible ability and dexterity, in razing this Bastille, and in gaining a free space, it is thus by no means intended at once to cover the site again and to encumber it with a new structure; we propose rather to make use of this area for the purpose of passing in review a pleasing and varied series of illustrative figures.

The third part is thus devoted to the historical account of early enquirers and investigators. As we before expressed the opinion that the history of an individual displays his character, so it may here be well affirmed that the history of science is science itself. We cannot clearly be aware of what we possess till we have the means of knowing what others possessed before us. We cannot really and honestly rejoice in the advantages of our own time if we know not how to appreciate the advantages of former periods. But it was impossible to write. Or even to prepare the way for a history of the theory of colours while the Newtonian theory existed; for no aristocratic presumption has ever looked down on those who were not of its order, with such intolerable arrogance as that betrayed by the Newtonian school in deciding on all that had been done in earlier times and on all that was done around it. With disgust and indignation we find Priestly, in his *History of Optics*, like many before and after him, dating the success of all researches into the world of colours from the epoch of a decomposed ray of light, or what pretended to be so; looking down with a supercilious air on the ancient and less modern inquirers, who, after all, had proceeded quietly on the right road, and who have transmitted to us observations and thoughts in detail which we can neither arrange better nor conceive more justly.

We have a right to expect from one who proposes to give the history of any science, that he inform us how the phenomena of which it treats were gradually known, and what was imagined, conjectured, assumed, or thought respecting them. To state all this in due connexion is by no means an easy task; need we say that to write a history at all is always a hazardous affair; with the most honest intention there is always a danger of being dishonest; for in such an undertaking, a writer tacitly announces at the outset that he means to place some things in light, others in shade. The author has, nevertheless, long derived pleasure from the prosecution of his task: but as it is the intention only that presents itself to the mind as a whole, while the execution is generally accomplished portion by portion, he is compelled to admit that instead of a history he furnishes the materials for one. These materials consist of translations, extracts, original and borrowed comments, hints, and notes; a collection, in short, which, if not answering all that is required, has at least the merit of having been made with earnestness and interest. Lastly, such materials, — not altogether untouched it is true, but still not exhausted, — may be more satisfactory to the reflecting reader in the state in which they are, as he can easily combine them according to his own judgement.

The third part, containing the history of the science, does not, however, thus conclude the subject: a fourth supplementary portion is added. This contains a recapitulation or revision; with a view to which, chiefly, the paragraphs are headed numerically. In the execution of a work of this kind some things may be forgotten, some are of necessity omitted, so as not to distract the attention, some can only be arrived at as corollaries, and others may require to be exemplified and verified: on all these accounts, postscripts, additions, and corrections are indispensable. This part contains, besides, some detached essays; for example, that on the atmospheric colours; for as these are introduced in the theory itself without any classification, they are here presented to the mind's eye at one view. Again, if this essay invites the reader to consult Nature herself, another is intended to recommend the artificial aids of science by circumstantially describing the apparatus which will in future be necessary to assist researches into the theory of colours.

In conclusion, it only remains to speak of the plates which are added at the end of the work; and here we confess we are reminded of that incompleteness and imperfection which the present undertaking has, in common with all others of its class; for as a good play can be in fact only half transmitted to writing, a great part of its effect depending on the scene, the personal qualities of the actor, the powers of his voice, the peculiarities of his gestures, and even the spirit and favourable humour of the spectator; so it is, in a still greater degree, with a book which treats of the appearances of Nature. To be enjoyed, to be turned to account, Nature herself must be present to the reader, either really, or by the help of a lively imagination. Indeed, the author should in such cases communicate his observations orally, exhibiting the phenomena he describes — as a text, in the first instance, — partly as they appear to us unsought, partly as they may be presented by contrivance to serve in particular illustration. Explanation and description could not then fail to produce a lively impression.

The plates which generally accompany works like the present are thus a most inadequate substitute for all this; a physical phenomenon exhibiting its effects on all sides is not to be arrested in lines nor denoted by a section. No one ever dreams of explaining chemical experiments with figures; yet it is customary in physical researches nearly allied to these, because the object is thus found to be in some degree answered. In many cases, however, such diagrams represent mere notions; they are symbolical resources, hieroglyphic modes of communication, which by degrees assume the place of the phenomena and of Nature herself, and thus rather hinder than promote true knowledge. In the present instance we could not dispense with plates, but we have endeavoured so to construct them that they may be confidently referred to for the explanation of the didactic and polemical portions. Some of these may even be considered as forming part of the apparatus before mentioned.

We now therefore refer the reader to the work itself; first, only repeating a request which many an author has already made in vain, and which the modern German reader, especially, so seldom grants:

Si quid novisti rectius istis
Candidus inperti; si non, his utere mecum!
(If you can better these principles, tell me in all honesty. If you cannot, join me in abiding by them!)

Johann Wolfgang von Goethe
Exposure of Newton's Theory

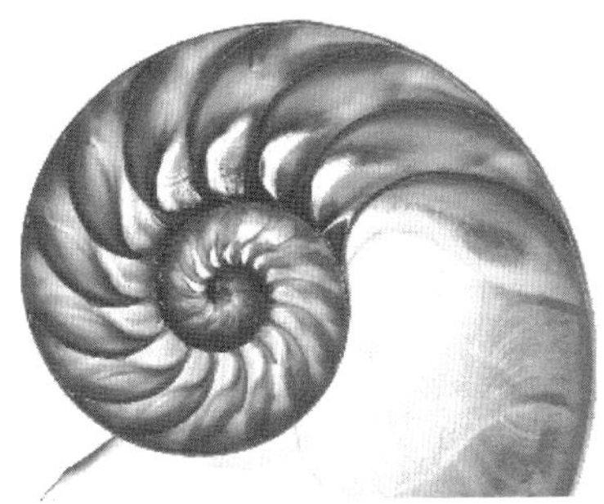

A typical spiral image of the kind favoured by the World Pantheist Movement

A Translation of Goethe's *Enthüllung Der Theorie Newtons*

By Michael Duck and Michael Petry

Annotated with particular reference to problems encountered
on translating both *Opticks* into German
and *Enthüllung* into English

Contents

Introduction   1
Digression   7

Newtonian Opticks, First Book, First Part   11

Proposition I. Theorem I.   13
Lights which differ in Colour, differ also in Degrees of Refrangibility

Proof by experiment   16
First experiment   18
Second experiment   23

Proposition II. Theorem II   35
The Light of the sun consists of Rays differently Refrangible

Third experiment   36
Fourth experiment   40
Fifth experiment   42
Sixth experiment   50
Seventh experiment   60
Eighth experiment   71
Recapitulation of the first eight experiments   77

Proposition III. Theorem III   **81**

*The sun's Light consists of Rays differing in reflexibility,
and those Rays are more reflexible than others which
are more refrangible*

Ninth experiment   82
Tenth experiment   86
Newton's Recapitulation of the first 10 experiments   86
Review of what follows   94

Proposition IV. Problem I   **97**

*To separate from one another the heterogeneous Rays
of compound Light*

Eleventh experiment   97

Proposition V. Theorem IV   **103**

*Homogeneal Light is refracted regularly without any Dilatation
splitting or shattering of the Rays, and the confused Vision of
Objects seen through refracting Bodies by heterogeneal Light
arises from the different Refrangibility of several sorts of Rays*

Twelfth experiment   104
Thirteenth experiment   105
Fourteenth experiment   107

Proposition VI. Theorem V   **113**

*The Sine of Incidence of every Ray considered apart, is to its
Sine of Refraction in a given Ratio*

Fifteenth experiment   118

Proposition VII. Theorem VI   **119**

*The Perfection of Telescopes is impeded by the different
Refrangibility of the Rays of Light*

Sixteenth experiment   122

Proposition VIII. Problem II — 125

To shorten telescopes

Newtonian Opticks, First Book, Second Part — 127

Proposition I. Theorem I — 131

*The phenomena of colours in refracted Light are not caused
by new modifications of the Light variously impress'd,
according to the various Terminations of the Light and Shadow*

First experiment — 131
Second experiment — 141
Third experiment — 142
Fourth experiment — 149

Proposition II. Theorem II — 151

*All homogeneal Light has its proper Colour answering to its
Degree of Refrangibility, and that Colour cannot be changed
by Reflections and Refractions*

Fifth experiment — 153
Sixth experiment — 157
Definition — 160

*Coloured Rays to speak properly are not coloured. In them there
is nothing else than a certain Power and Disposition to stir up a
Sensation of this or that Colour*

Proposition III. Problem I — 163

*To define the Refrangibility of the several sorts of homogeneal
Lights answering to the several Colours*

Seventh experiment — 163
Eighth experiment — 165

Proposition IV. Theorem III

*Colours may be produced by Composition which shall be like
to the Colours of homogeneal Light as to the Appearance
of Colour, but not as to the Immutability of Colour and
Constitution of Light. And those Colours by how much they
are more compounded by so much are they less full and intense,
and by too much Composition they may be diluted and
weaken'd till they cease, and the Mixture becomes white or grey.
They may also be Colours produced by Composition, which
are not fully like any of the Colours of homogeneal Light*

171

Proposition V. Theorem IV

*Whiteness and all grey Colours between white and black,
may be compounded of Colours, and the whiteness of the sun's
Light is compounded of all the primary Colours mix'd in
a due Proportion*

177

Ninth experiment 177
Twelfth experiment 180
Eleventh experiment 186
Tenth experiment 187
Analysis of the tenth experiment 187
Thirteenth experiment 188
Fourteenth experiment 191
Fifteenth experiment 192

Proposition VI. Problem II

*In a mixture of Primary Colours, the Quantity and Quality
of each being given, to know, the Colour of the Compound*

201

Proposition VII. Theorem V

*All the Colours in the Universe which are made by Light,
and depend not on the Power of Imagination, are either the
Colours of homogeneal Lights, or compounded of these, and that
either accurately or very nearly, according to the Rule of the
foregoing Problem*

203

Proposition VIII. Problem III **205**

*By the discovered Properties of Light to explain the Colours
made by Prisms*

Sixteenth experiment 208

Proposition IX. Problem IV **211**

*By the discovered Properties of Light to explain the Colours
of the Rainbow*

Proposition X. Problem V **213**

*By the discovered Properties of Light to explain the permanent
Colours of Natural Bodies*

Seventeenth experiment 214

Proposition XI. Problem VI **229**

*By mixing colour'd Lights to compound a beam of Light
of the same Colour and Nature with a beam of the sun's
direct Light, and therein to experience the truth of the
foregoing Propositions*

Conclusion **231**
In Lieu of an Epilogue **233**

Introduction

1.

Although in the first part we have kept as much as possible to the didactic approach and avoided any actual polemics, it was inevitable that at certain junctures various misgivings should have arisen concerning the established and prevailing theory. What is more, this outline of our doctrine of colours is by its very nature polemical, for we have attempted to provide a comprehensive collation of the phenomena and to order them so that all may be obliged to consider them in their proper sequence and actual inter-relationships, and so that in future those who are in fact merely involved in dwelling upon particular appearances[1] in order to lend credence to their hypothetical assertions, may find their handiwork more difficult.

---

[1] Goethe's word, *Erscheinungen* has been translated here as "appearances". A technical translator might be tempted to treat the word as a synonym of *Phänomene* and render it as "phenomena", but this would be incorrect. For instance, in paragraph 59: *(Die) bringen dem, der mit dieser Erscheinung nicht bekannt ist, eine ganz besondere Konfusion in das Phänomen* (Anyone who is not familiar with this appearance will find the phenomenon particularly confusing).

In other places he might prefer "effect" (e.g., in paragraph 311: *... jene Randerscheinungen ...*), or "event" (e.g., in paragraph 74: *... zufällige Erscheinungen ...*) but this would be to ignore the significance of the word from a philosophical standpoint. Goethe was more interested in the perception of colour than in the physics of it. See 'Appearance and Reality' in *The Encyclopaedia of Philosophy*, edited by P. Edwards (New York, 1967) and 'Erscheinung' in *Historisches Wörterbuch der Philosophie*, edited by J. Ritter (Darmstadt, 1972).

2.

Although there has, hitherto, been a general belief that the nature of colour was understood and people have imagined that it was being accounted for by means of a trustworthy theory, this was by no means the case, for it was hypotheses that were being accepted as fundamental, which were giving rise to a contrived ordering of the phenomena and which generated a peculiar and jejune doctrine, promulgated with an extraordinary boldness.

3.

How the founder of this school, the remarkable Newton, arrived at such a preconception, how he established it and the various means by which he communicated it to others, will become evident in the *History*. Here, we are concerned with the work he published under the title of *Opticks*, in which he finally set down his convictions, that is to say, reassembled and rearranged his earlier writings. He published this work late in life, and he himself declared it to be a complete statement of his convictions, he refused to delete or to add a single word and arranged for the Latin translation to be made under his personal supervision.

4.

The earnestness with which this task was undertaken, the attention to detail in the way in which it was carried out, aroused the greatest confidence. It gradually became a general conviction that the book contained what was irrefutably true, and people still regard it as a masterpiece of the scientific treatment of natural phenomena.

5.

It is, therefore, useful and necessary for our purpose to translate and comment upon certain excerpts from this work, in order that those who go on to concern themselves with the subject may be provided with a thread for finding their way through such a labyrinth. Before we begin with the business itself, however, a few introductory remarks are called for.

6.

It goes without saying that when a teacher deals with the things of Nature he has the choice of either proceeding from experiences to principles or from principles to experiences; it is also acceptable, moreover, and may even be necessary, that he should switch from one procedure to the other. Newton makes unwarranted use of this dual method in advocating his own cause, however, for after starting off by accepting as already known what ought first to be introduced, derived, explained, and demonstrated, he seeks out from the great mass of material only those phenomena which appear to add plausibility to what has already been stated, and we are, therefore, obliged to demonstrate and specify the unsystematic and wilful manner in which he conducts these experiments, of which he never gives a straightforward account, and which he simply continues to duplicate and pile one upon another, until even the most capable head prefers to acknowledge the credibility of the chaos, rather than commit himself to the interminable task of reconciling and ordering these belligerent elements. It would have been quite impossible for us to do so had we not previously paid careful attention to sorting out chromatic phenomena into a certain natural connectedness and thus been able to bring out more clearly the artificial and arbitrary way in which they had been arranged and disarranged. We are now able to make immediate reference to a natural exposition, and therefore to cast a wholesome light on this excess of confusion and complication. This is the only way in which it will be possible to settle an argument which has already been going on for a century, and which, whenever it has been re-opened, has been quashed and dismissed by the predominant school of thought as foolhardy, presumptuous, or even ludicrous and absurd.

7.

How it was that obstinacy of this kind was possible will gradually become apparent to our readers. By employing an artificial method, Newton had imparted to his work such an appearance of rigour that it was admired by those who appreciated its form and regarded as astonishing by the layman. The venerable guise of a mathematical treatment was added, and was used by him to give credence to the whole.

8.

Thus, the work begins with definitions and axioms, which will be examined later, when they will no longer be able to impress our readers. We are then confronted with propositions, which repeatedly and unvaryingly assert what has yet to be proved; theorems, which give expression to things no one is able to observe; experiments, which under various conditions invariably turn out to be duplications and which involve an excessive preoccupation with the same extremely limited context; problems, moreover, which are insoluble — all of which is to be demonstrated in detail in what now follows.

9.

In English, the title of the work is; *Opticks, or a Treatise of the Reflections, Refractions, Inflections, and Colours of Light*. Although the English word *Opticks* may have a somewhat more limited connotation than the Latin *Optice* or the German *Optik*, there can be no doubt that it implies more than is dealt with in this treatise. The work deals exclusively with colour, with the appearances of colour. Everything else relevant to natural and artificial perception is all but excluded and, comparing it in this respect with the *Optical Lectures*, one will find in it nothing of the great mass of genuinely mathematical matters contained in these lectures.

10.

It is necessary that this should be observed here at the outset; for even the title of the treatise has given rise to the misconception that its subject matter and execution are mathematical, whereas the subject matter is merely physical and the mathematical treatment merely apparent; what is more, the advance of science has long since made it obvious that, since Newton the physicist failed to carry out his observations with precision, the formulae by means of which he gives expression to his experimentation have to be regarded as inadequate and erroneous; which can be confirmed in detail by consulting any account of the discovery of the achromatic telescope.

11.

This so-called *Opticks*, or more properly *Chromatics*, consists of three books; the first of these is divided into two sections, and it is this book alone which is now to be treated polemically. We have usually translated from the fourth edition of the English original, London 1730, which is written in a natural, simple style. The Latin translation is extremely close and precise, but on account of the Roman manner of speech somewhat more pompous and dogmatic.

12.

Since we only cite extracts, however, we saw no point in having all Newton's plates re-engraved, we are compelled to make frequent reference to the work itself, which those of our readers who are genuinely interested in the matter will in any case have at their side, earlier in the original or in translation.

13.

We have had the passages which have been translated directly, in which the opponent himself speaks, set in a smaller type, and our observations set in the larger type with which our readers are already familiar.

14.

We have, moreover, numbered the paragraphs into which our work could be divided. As in the *Outline of a Theory of Colours*, we have done this here, not in order to make the work appear more consistent, but simply in order to facilitate referencing and cross-referencing, which can be a help to both supporters and opponents. When we quote the *Outline*, the paragraph number will be preceded by an O.

Digression

15.

The greater part of what precedes and what follows was written when the question arose as to whether it was advisable to state briefly here at the outset the difference between the view we hold and that which originates from Newton and which has spread throughout the learned and the uneducated world alike.

16.

We begin by observing that the manner of thinking of which we approve is neither anything peculiar to ourselves, nor is it put forward as a novel doctrine, hitherto unknown. Clear traces of it are to be found in former times, in spite of all changes of opinion it has survived the centuries, and has been reexpressed time and again, as will become more fully apparent from the *History*.

17.

Newton maintains that white colourless light everywhere, but particularly the light of the sun, actually contains different lights of various colours (in that they stimulate the sensation of colour), the mixing of which produces white light (the sensation of white light).

18.

In order to make these lights appear, however, he submits white light to various conditions — transparent bodies which deflect light from its path, opaque bodies which reflect it, others which it impinges upon; but these conditions are not enough for him. He makes use of various forms of refracting media, he arranges the room in which he is operating in various ways. He restricts the light by means of small apertures, minute slits, and attenuates it in a hundred different ways. He then asserts that all these conditions influence it only in so far as they excite its properties or facilities (fits), eliciting thereby its inner nature, and bringing forth what is hidden within it.

19.

These coloured lights are the integrating components of his white light. Nothing is added to light by all the operations mentioned above, nothing is taken away, all that happens is that there is a revealing of its capabilities or content. If it displays various colours in refraction, it is diversely refrangible; if it displays colours in reflection it is diversely reflexible, etc. Each new appearance indicates a new capability on the part of light in revealing itself, yielding up its content.

20.

The doctrine to which we subscribe, however, and which we only mention here in so far as it is opposed to that of Newton, is also concerned with white light. It, too, makes use of external conditions in order to produce coloured appearances. It ascribes significance and value to these conditions, however, and rather than imagining that it can develop colours out of light, attempts to make the point that it is light, together with that by which it is restricted, which gives rise to colour.

21.

For even if we confine ourselves to the case of refraction, which is Newton's main preoccupation in the *Opticks*, it is certainly not by refraction that the colours are elicited from light; there is a further essential condition — the refraction has to affect and displace an image. An image occurs only where there are boundaries; Newton completely overlooks these boundaries and even denies their significance. In this case, however, we ascribe to the image[2] as well as to its surroundings, to the bright centre as well as the dark boundary, to the activity as well as to the limit, the same overall effect. All the experiments lend support to our doing so, and the more we diversify them the more they bear out our affirmation, the plainer and clearer the matter becomes. We proceed from what is simple whilst postulating an interdependent opposition, and by means of its combination call forth the world of colour.

22.

Newton appears to proceed from what is simpler, for he attempts to deal merely with light; but although like us he also confronts light with conditions, he denies these conditions their integrating role in what is brought forth. His theory merely has the appearance of being monadic or unitary. He introduces into the unity of this appearance the diversity he wants to derive from it, whereas we believe it to be far better to develop and construct this diversity from the avowed duality.

23.

We now have to undertake the proper task of this polemical section, which is to show how he sets about making what is false true and what it true false.

---

[2] Goethe's word *Bild* occurs frequently in the text. D.E. Miller has suggested that this word is sometimes best translated as "image" but at other times "form" or "object"; depending upon the precise meaning Goethe is attempting to convey; see D.E. Miller, '*Die Farbenlehre* in English' in *Goethe and the Sciences: A Reappraisal.* F. Amrine, F. Zucker, and H. Wheeler (eds), Reidel (Dordrecht, 1987) pp. 101–112 (p. 108). The word is an important one for it is inextricably linked to Goethe's philosophical approach to light. Here, refraction is being discussed. Whereas Newton would talk of rays being bent, Goethe sees images being *verrückt* [displaced]. This is in keeping with his views on the role of the observer, a ray exists quite independently of the observer and is not perceivable, being merely a concept to aid a hypothesis. An image, on the other hand, requires an observer. Furthermore, a ray can be considered without reference to its surroundings, whereas an image depends upon its outline — the light/dark interface; another reason for Goethe's preference for *Bild.*

Newtonian Opticks
First Book, First Part

Proposition I. Theorem I

Lights which differ in Colour, differ also in Degrees of Refrangibility

24.

If we freely accept, right at the start, that the treatise with which we are concerned is a wholly consistent entity, we may also observe, that in the opening words quoted above, in this proposition which confronts us at the outset, the whole doctrine is present in a nutshell, so that even here, a captious method by means of which the author misleads us throughout the whole of the book, is already in full operation. In order to demonstrate that this is the case, to make it plain and obvious, we must be careful not to let him get away with a single word, a single phrase; and we urge our readers to pay the utmost attention to the matter, that they may feel themselves to be permanently free from bondage to this doctrine.

25.

Lights — The surreption and circumvention employed by Newton with such insidiousness throughout the entire treatise is already fully apparent in this plural. Lights, several lights! And what sort of lights?

Which differ in colour — In the first and second experiments, which are supposed to prove the matter, coloured papers are produced, and without further ado the effects they transmit to our eyes are treated as lights. Quite evidently a hypothetical expression for commonsense simply tells us that light acquaints us with the various properties of surfaces; it is not to be assumed that that which reflects from them can be regarded as a heterogeneous light.

In any case, we are already on to coloured lights before any mention has been made of an achromatic light. We are already dealing with coloured lights, and only later are we to be told what the why and wherefore of their origin may be. Yet anyone who has weighed the evidence adduced in the *Outline* of our *Theory of Colours*, will be convinced that one cannot discuss lights at this juncture. For we have amply demonstrated that the determinate being of all colours depends upon a light and a non-light, that colour constantly tends towards darkness, that it is a *σχιερόν*,[3] that when in any way whatsoever we project a colour on to a light object, we are not illuminating but darkening it. Since Newton begins his whole exposition in this extremely contrived manner, by considering shadowed light and semi-darkness, it is not surprising that he should manage to keep those who accept his initial proposition in a state of benightedness and semi-darkness.

26.

Also in refrangibility — How this abstract word suddenly pops up! Since it has, however, already been used in the axioms, the attentive and credulous disciple is already thunderstruck by these miracles, and is no longer free to investigate critically what he is being told.

27.

Differ — It is by means of refrangibility, therefore, that we are initiated into a great mystery. Light, this essence, which we are aware of only as a unity, as indivisible, is now presented to us as a composite, as consisting of a variety of parts, as something which acts in a diverse manner.

---

[3] Goethe did not consider darkness to be merely an absence of light, but something positive. He probably used the Greek word for it so that it would stand out in the text as representing something not to be ignored.

We readily admit that diversity can develop from or within a unity, that there can be differentiation; there are, however, various ways in which this can take place. Here, we shall only make mention of two: firstly, that in which an opposition emerges whereby the unity manifests itself as having two aspects and so becomes capable of more pronounced effects; secondly, that in which the development of the differentiation proceeds steadily as a series. Newton is totally unaware that the first one can occur with prismatic appearances, although the phenomenon provides him with enough opportunities to adopt this manner of interpretation. He does not hesitate, however, to plump for the second one. The refrangibility is not only diverse, but also operative.

28.

In Degrees — And immediately there is, therefore, a sequential and separated image, a scale, a complete image consisting of a variety and yet of an infinity of parts, which is confluent and yet separable and at the same time inseparable, a spectre, which for a century has been able to hold the scientific world in awe.

29.

If this proposition were to express anything which accords with experience, it might perhaps be so formulated: images which differ in colour appear to be displaced by refraction in various ways. Had one used this manner of expression one could, if need be, thus describe the phenomenon of the first experiment. The appearance could be called a diverse refraction, and one could then investigate more precisely what it actually looks like. To pitch us straight into the -ibilities and -tities[4] to expect us to be willing not only to accept the way in which they are proved but also to concern ourselves with them alone in order to obtain proof, is to ask too much of us.

---

[4] *daß wir sogleich zu den Ibilitäten. zu den Keiten geführt worden.* Goethe was scornful about the use of such words as refrangibility, reflexibility etc. — which imply that light possesses a certain versatility — a quality quite out of tune with his own theory.

Proof by experiment

30.

We have no wish to begin by frightening our readers with some sort of paradox; but we cannot refrain from maintaining that nothing can actually be proved by experiences and experiments. Phenomena may be observed with great precision, experiments may be well prepared, one can bring experiences and experiments into a certain order, one can deduce one appearance from another, one can present a certain order, one can present a certain field of knowledge, one can raise one's views to certainty and completeness, and that, I should have thought, would be sufficient. Each will draw his own conclusions from this, however, nothing is to be proved by it, certainly no -ibilities and –tities, all opinions concerning things are the individual's, and we know only too well that conviction arises from the will not from insight, that no one comprehends anything, unless it accords with him, and he is therefore able to admit it. In knowledge as in action everything is determined by prejudice and, as is evident from the word itself, prejudice is a judgement which precedes investigation. It is an affirmation or denial of that which appeals to or contradicts our nature; it is a joyous impulse on the part of our living being toward what is true or toward what is false, toward everything with which we feel in harmony.

31.

We do not, therefore, imagine we can prove Newton to have been wrong; for anyone who thinks in atomistic terms, anyone who keeps to what is traditional, anyone who defers in solemn awe to a traditionally venerated name, anyone who wants a quiet life, will much prefer to repeat Newton's first proposition, swear it to be true, give assurance that it has all been demonstrated and proved, and execrate our efforts.

Indeed, we readily admit that over the years we have often been reluctant to reopen this business. To disturb the hallowed conviction of the Newtonian school, to introduce uneasiness into the divine tranquillity with which the whole semi-educated world accepts it as creditworthy, might well be regarded as a sin. For when all who profess the old inflexible doctrine constantly re-iterate it in the lecture rooms, the students readily adopt the short formulae by means of which the whole thing is dealt with and disposed of, and the public at large absorbs the divine dogma out of the air, as it were; in this connection, I cannot refrain from repeating the anecdote concerning the fortunate fellow who, after hearing something of the more recent developments, gave an assurance that Newton had said it all before and much better; although he did not quite know where.

32.

In that we now turn to the experiments, we ask our readers to be fully attentive right from the start, for the author sets off with a breakneck leap and without warning hurls us into the middle of the matter, so that if we are not on our guard we shall be taken by surprise, become confused, and lose the ability to judge freely.

33.

Those friends of science who are sufficiently acquainted with the subjective dioptric experiments of the second class, which have been expounded and deduced by us in sufficient detail, will realise immediately that Newton's procedure here is certainly not mathematical. For when the mathematician wishes to instruct, he presupposes what is most simple, and constructs his wonderful edifice out of those elements which are most comprehensible. Newton, however, starts off with what is perhaps the most complicated subjective experiment, conceals its origin, avoids exhibiting its various aspects, and surprises the unwary student who, once he has given his approval, falls into this trap without knowing how to get out of it.

On the other hand, anyone who is aware of the proper context of this first experiment will have no difficulty in avoiding the other fetters and chains, and will take pleasure in casting them off if they have already been fastened upon him by tradition.

First experiment

34.

I took a black oblong stiff Paper terminated by Parallel Sides, and with a Perpendicular right line drawn across from one Side to the other, distinguished it into two equal Parts. One of these parts I painted with a red colour and the other with a blue. The Paper was very black, and the Colours intense and thickly laid on, that the Phenomenon might be more conspicuous.

35.

It is an entirely unnecessary condition that the paper here should be black. For if the blue and red are laid on thickly enough, the base can no longer show through, whatever its colour, if one is familiar with the Newtonian hypothesis, however, one sees roughly what the reason for this must be. He makes use of a black base so that it cannot be suggested that anything of his supposedly unresolved light is transmitted through the colours that have been laid on. As has already been observed, however, given the situation it is an entirely useless condition, and nothing is more of a hindrance to genuine insight into a phenomenon or an experiment than superfluous conditions. In fact all that needs to be said is that one takes two equal rectangular pieces of stiff paper, one red, the other blue, and places them carefully side-by-side.

In order to give a proper account of his experiment, the author should have gone on, first and foremost, to specify the precise locality, that is to say, the position and placing of this two-coloured paper, instead of leaving it to the reader to pick things up here and there as best he can from what follows, and so run the risk of a misunderstanding.

36.

This Paper I view'd through a Prism of solid Glass, whose two Sides through which the Light passed to the Eye were plane and well-polished, and contained an Angle of about 60 degrees; which Angle I call the refracting Angle of the Prism. And whilst I view'd it I held it and the Prism before a Window in such a manner that the Sides of the Paper were parallel to the Prism, and both those sides and the Prism were parallel to the Horizon, and the cross-line, which divided the two colours, was at right angles to it.[5]

37.

The word *parallel* in the English text is quite clearly a misprint, and should read *at right angles.* For the long sides of the coloured paper and the cross-line cannot both be parallel to the horizon. The Latin text has *perpendicular,* which is perfectly correct; nevertheless, since what is being discussed is not a ground plan but a spatial relationship, it could easily be taken to mean vertical, which would confuse the experiment. For the coloured paper must he flat, and the short sides must, as we have said, make a right angle with the horizon or, if preferred, with the windowsill.

38.

And that the Light which fell from the Window upon the Paper made an Angle with the Paper, equal to that Angle which was made with the same Paper by the Light reflected from it to the Eye.

39.

How can ordinary daylight — for it does not look as though sunlight is being discussed — which falls upon the paper from all directions, be said to make an angle with it? It is in any case a totally superfluous condition; for the arrangement could just as well be set up beside the window.

40.

Beyond the Prism was the Wall of the Chamber under the Window covered over with black Cloth, and the Cloth was involved in Darkness that

---

[5] See Figure 11 in the Appendix.

*no Light might be reflected from thence, which in passing by the Edges of
the Paper to the Eye might mingle itself with the Light of the Paper, and
obscure the Phenomenon thereof.*

41.

Why did he not say: beyond the coloured paper? That is certainly
nearer the window, and the purpose of the black cloth is simply to
provide the coloured paper with a dark background. An exact and
clear account of this arrangement would therefore be as follows: the
wall is covered with black cloth from the windowsill to the floor.
A cardboard parallelogram is taken, half of it is covered with red paper
and the other half with blue paper, so that these two halves meet at
the shorter cross-line. This cardboard is placed flat, at about half the
height of the darkened parapet, so that when it is observed by someone standing further away from the window it appears against a black
background, and nothing can be seen of the stand upon which it is
resting. Its longer sides should be parallel to the window wall, and the
prism through which the observer views the paper in question is also
held in this position, initially with its refracting angle turned upwards,
subsequently with it turned downwards.

Does this elaborate description signify anything more than that
the two-coloured paper in question is placed against a black background, or that a red and a blue quadrilateral are pasted horizontally
and side-by-side on to a black surface, and then positioned in front of
the observer? For it makes no difference at all if the black base is illuminated to a certain extent and appears as a dark grey; the phenomenon remains the same. Such pedantic exactitude in excluding all the
unresolved light postulated in his hypothesis, in order to endow his
experiments with a kind of purity, is exhibited throughout Newton's
investigations. As we shall make abundantly clear, it is quite pointless,
and gives rise to nothing but useless requirements and conditions.

42.

*These Things being thus ordered, I found that if the refracting Angle of
the Prism be turned upwards, so that the Paper may seem to be lifted*

upwards by the Refraction, its blue half will be lifted higher by the Refraction than its red half. But if the refracting Angle of the Prism be turned downward, so that the Paper may seem to be carried lower by the Refraction, its blue half will be carried something lower thereby than its red half.

43.

In our *Outline of the Theory of Colours* we have given a sufficiently detailed account of the second class of dioptric colours and particularly of the subjective experiments, paying special attention in Chapter 18 (O. 258–284) to showing precisely what is actually involved when coloured images are displaced by refraction. It is made perfectly plain there, that coloured fringes form on coloured and uncoloured images and that they are either homonymous or heteronymous with the surface, heightening its colour in the first instance, and smudging and obscuring it in the second. It is this which escapes the notice of an inattentive or blindly prejudiced observer, and which lead our author to draw the following rash conclusion:

44.

Wherefore in both Cases the Light which comes from the blue half of the Paper through the Prism to the Eye, does in like Circumstances suffer a greater Refraction than the Light which comes from the red half and by consequence is more refrangible.

45.

This is the foundation and cornerstone of Newton's optical work; it is the experiment examined here which he took to be so important that he selected out of hundreds and gave it precedence over all other chromatic experiences. We have already noted (O. 268) the captiousness and juggling with which this experiment was carried out; for if the appearance is to be at all deceptive, the red has to be a vermillion and the blue a very dark blue. One is immediately aware of the illusion if one uses light-blue. And why has it never occurred to anyone to go on to ask another embarrassing question? According to the

Newtonian doctrine, yellow–red is the least refrangible colour, violet the most; why, then, does he place a dark blue and not a violet paper next to the red? If the theory were correct, the difference in refrangibility between yellow–red and violet would be much greater than that between yellow–red and blue. But the fact is, of course, that a violet paper conceals the prismatic fringes less than a dark blue one, as has been made perfectly clear to any observer by our presentation of the matter. Anyone who uses his eyes and his intellect will, however, be impressed by Newton's powers of observation and the precision of his experiments; one might even go on to ask who could have imposed such jugglery upon a person as extraordinarily gifted as Newton was, had he not done so himself? Only if one recognises and understands the power of self-deception, the extent to which it verges upon dishonesty, is one able to reach an understanding of the procedures of Newton and his school.

46.

We now wish to call attention to Newton's figure, the eleventh on the second of his plates, for it certainly needs to be examined. It is drawn in a confusing manner with regard to perspective and is misleading in another odd way. The two-coloured cardboard is distinctive in that one half is dark and the other light, and the perpendicular orientation of its surface with respect to the window is made fairly clear; however, the eye, together with its prism, is not in the right position; it ought to be *in line with* the cross-line of the coloured cardboard. But the displacement of the images is badly drawn; it looks as though they are displaced diagonally, which is not the case; for they are either displaced towards or away from the observer, depending on the position of the refracting angle. No one, however, ought to overlook the oddest thing of all. The displaced images, which according to the Newtonian theory are diversely refracted as shown in the original edition as having borders, those in the dark half being indistinct but those in the light half being very clear-cut, as they are also in the plates of the Latin translation. If all that happens in this experiment is that one image is displaced more than the other, why does he not represent the images as still bounded by their lines, why does he make them broader, why does he give them fuzzy borders? These borders did not escape his notice, then — although he could not convince himself that it was they and not a diverse refrangibility which gave rise to the phenomenon. Why, therefore, does he not mention in the text these appearances he had so painstakingly, if not quite correctly, etched into the copper? A Newtonian will probably reply that it is the ineradicable factor of undecomposed light which creates this irregularity.

Second experiment

47.

We now have to show the extent to which this experiment resembles the former one in also being based upon an illusion. In this case, however, we consider it to be more appropriate not to interrupt the author, but to hear him out before responding.

48.

About the aforesaid Paper, whose two halves were painted over with red and blue, and which was stiff like thin Pasteboard, I lapped several times a slender Thread of very black Silk, in such a manner that several parts of the Thread might appear upon the Colours like so many black Lines drawn over them, or like long and slender dark Shadows cast upon them. I might have drawn black Lines with a Pen, but the Threads were smaller and better defined.

49.

This Paper thus coloured and lined I set against a Wall (perpendicularly to the Horizon)[6] so that one of the Colours might stand to the Right Hand, and the other to the Left. Close before the Paper, at the Confine of the Colours below, I placed a Candle to illuminate the Paper strongly: for the Experiment was tried in the Night.[7]

---

[6] This phrase was omitted by Goethe.
[7] See Figure 12 in the Appendix.

50.

The Flame of the Candle reached up to the lower edge of the Paper, or a very little higher. Then at the distance of six Feet, and one or two Inches from the Paper upon the Floor I erected a Glass Lens four Inches and a quarter broad, which might collect the Rays coming from the several Points of the Paper, and make them converge towards so many other Points at the same distance of six Feet, and one or two Inches on the other side of the Lens, and so form the Image of the coloured Paper upon a white Paper placed there, after the same manner that a Lens at a Hole in a Window casts the Images of Objects abroad upon a Sheet of white Paper in a dark Room.

51.

The aforesaid white Paper, erected perpendicular to the Horizon, and to the Rays which fell upon it from the Lens I moved sometimes towards the Lens, sometimes from it to find the places where the Images of the blue and red Parts of the coloured Paper appeared most distinct. Those Places I easily knew by the Images of the black Lines which I had made by winding the Silk about the Paper. For the Images of those fine and slender Lines (which by reason of their Blackness were like Shadows on the Colours) were confused and scarce visible, unless when the Colours on either side of each Line were terminated most distinctly. Noting therefore, as diligently as I could, the Places where the Images of the red and blue halves of the coloured Paper appeared most distinct I found that where the red half of the Paper appeared distinct, the blue half appeared confused, so that the black lines drawn upon it could scarce be seen; and on the contrary, where the blue half appeared most distinct, the red appeared confused, so that the black Lines upon it were scarce visible. And between the two Places where these Images appeared distinct there was a distance of an Inch and a half; the distance of the white Paper from the Lens when the Image of the red half of the coloured Paper appeared most distinct, being greater by an Inch and a half than the distance of the same white Paper from the Lens, when the Image of the blue half appeared most distinct. In like Incidences therefore of the blue and red upon the Lens, the blue was refracted more by the Lens than the red, so as to converge sooner by an Inch and a half, and therefore is more refrangible.

52.

Now that we have heard the author out, familiarised ourselves with his apparatus and with what he believes he can do with it, we shall put forward our observations on this experiment under various headings and attempt to resolve it into its various elements, this being the most profitable approach whenever Newton is being tackled.

53.

Our observations are therefore concerned with: (1) the experimental situation, (2) the illumination, (3) the lens, (4) the image produced, and (5) the conclusion drawn from the appearances.

54.

(1) *The experimental situation.* Before we make any further use of the two-coloured cardboard we encountered in the experiment in question, let us examine it and its properties rather more closely.

55.

When paper the colour of red lead is placed alongside paper which is dark blue, the former appears to be light, the latter dark and, particularly at night, almost black. If black threads are then wound around both, or if black lines are drawn upon them, it is obvious that at a distance at which the naked eye is able to pick out the black lines on the bright red it is still unable to distinguish them on the blue. Take two men, one in a scarlet jacket, the other in a dark-blue jacket, both garments having black buttons; if one gets them to walk side-by-side along the street toward the observer, the buttons on the red jacket will be seen much sooner than those on the blue one, and both persons will have to be quite near before the eye perceives both garments and their buttons with equal distinctness.

56.

If this experiment is to be seen in its proper context it has, therefore, to be diversified. A rectangular surface is divided into four equal squares, each of which is given a particular colour. Black lines are drawn across all of them, and if they are observed from a certain distance with the naked eye or with eye-glasses and the distance is altered, it will certainly be discovered that sooner or later the black threads become apparent to the sense of vision, not because the different coloured backgrounds possess particular properties, but simply in so far as one is brighter than the other. In order that all doubt may be removed, white threads are wound around the various coloured papers, or white lines are drawn across them, reversing the situation. In order to be fully convinced, one dispenses with the colours entirely and repeats the experiment with white, black, and grey papers; one cannot then fail to see that the difference in clarity is merely a consequence of the degree of contrast between light and dark. And this is quite evidently also the case in the experiment prescribed by Newton.

57.

(2) *The illumination.* The constructed image can be very brightly illuminated by a row of burning candles hidden from the lens, or by three candles so grouped that their several wicks appear to constitute one flame. These are then screened from the lens and, during observation, an assistant moves them gently to and fro very close to the constructed image so that all parts of it are brightly lit in succession. For a very powerful illumination is required if the experiment is to be at all effective.

58.

(3) *The lens.* At this juncture, we find we have to make some general preliminary observations if what is to be dealt with here and later is to be properly understood.

59.

Each image is displayed on an opposing smooth surface, where its linear form is apparent. It is also observable on a rough surface if the separate parts of the image are exclusively reflected by separate parts of the corresponding surface. External objects can be projected on to a white screen inversely through the small aperture of a *camera obscura*.

60.

The interspace through which this imaging takes place is considered to be empty; the images are displaced by that which fills the space, despite it being transparent. It is the phenomena which arise when media displace images, and especially coloured appearances, which are of particular interest to us here.

61.

We are particularly concerned with the operations brought about by prisms with a three-sided base and by lenses.

62.

Lenses are, in effect, an infinite array of prisms; convex ones being an array of prisms standing back-to-back and concave ones being an array of prisms standing apex-to-apex, and in both cases assembled about a centre with curvilinear surfaces.

63.

The standard prism, with its refracting angle turned downwards, draws objects towards the observer; when the prism's refracting angle points upwards the objects are pushed away from the observer. When both operations are considered with respect to one's surroundings, the first reduces the space around the observer, the second extends it. Considered objectively, therefore, a convex glass must enlarge and a concave one diminish; in what we call the objective operation, therefore, the opposite occurs.

64.

The convex lens, which is what we are actually using here, causes the images which pass through it to contract. The most important image is that of the sun. If this image is allowed to pass through the lens and be displayed just behind it on a screen then, as the screen is moved backwards, one will see the image diminish until it reaches a position in which its size is at a minimum relative to the lens and it is to be seen at its clearest.

65.

Even earlier on in these experiments considerable heat was generated and the screen was burnt, especially if it was black. Although this effect occurs beyond the sun's image point as readily as it does prior to it, the image point can be said to coincide with the most powerful burning point.

66.

The sun is the most distant image which can be reproduced during daytime. That is why it is also the first to be concentrated into a firm and precise delineation by the action of the lens. If one wishes to observe the clouds clearly, the screen has to be moved further from the lens. Hills and woods, houses, the nearby trees, all eventually manifest themselves, one after the other, and the sun's image has already expanded considerably beyond its image point, before these nearer objects reach theirs. This much is known from experience where the representation of external objects through lenses is concerned.

67.

In the experiment we are interpreting here, the differently coloured surfaces which are to be projected behind the lens together with their black threads, are placed side-by-side. Consequently, if one shows up distinctly before the other, the reason for this is not to be sought in any difference in the distances.

68.

It is from this that Newton wants to derive proof for his diverse refrangibility; as we have now shown by means of our consideration of the case in question, the actual cause of the various appearances behind the lens is simply the varying distinctness of the images projected on to the differently coloured backgrounds. We now have to give a more precise demonstration that this is the case.

69.

We shall first describe the apparatus we setup in order to perform the experiment properly. At one end of a horizontal stand there is an attachment into which the object may be inserted. The lights can be placed in front of it in a recess. The lens is fixed in a vertical board which can be moved backwards and forwards along the stand. Within the stand there is a moveable frame, at the end of which is erected a screen onto which the image is projected. By this means, the lens can be moved toward either the object or the screen, and the screen moved either toward or away from both, and the three components; object, lens, and screen, are exactly parallel to each other. Once one has found the position suitable for observation, the frame can be secured with a screw. This apparatus is easy to use and reliable, since everything is deployed in precisely adjustable positions. One now searches for the point at which the image is most distinct by moving the lens and the screen to and fro. When it has been found, observation can begin.

70.

(4) *The image produced.* As is Newton's wont, he pitches us into the middle of the matter with his bright red and blue pasteboard. It has already been observed that the experimental situation has first to be diversified and investigated in order to find out what can be expected of the image. We therefore approach it as follows: we place four squares together on a pasteboard in the form of a larger square, their shades being black, white, and dark and light grey. We draw black and white lines across them, and with the naked eye note the extent to which they are distinguishable against the various backgrounds. Now, since Newton refers to his black threads as images, why does he not perform the experiment with actual small images? We therefore place small light and dark images such as the diamonds or discs or figures found on playing cards on the four squares described above, and make use of this embellished pasteboard as our subject. Only now are we able to undertake a reliable investigation of what is to be expected of the image.

71.

Any image illuminated by candles shows up less clearly than it would in sunlight; yet an image so illuminated and, what is more, after passing through a lens, is supposed to yield an image which is distinct enough to constitute the basis of a significant theory.

72.

If we now illuminate our above-mentioned pasteboard as strongly as possible and attempt by means of the lens to project its image as distinctly as possible on to the white screen, all we shall ever get is a blurred reproduction of it. Black appears as a dark grey, white as a light grey; the dark and light grey of the pasteboard are less distinguishable than they are to the naked eye. It is precisely the same with the images. Those which contrast most sharply in lightness and darkness are also the most distinct. Black on white, white on black, may be clearly distinguished; white and black on grey is already somewhat less distinct, although it still displays a certain degree of clarity.

73.

If we now prepare an arrangement of coloured squares, we have to bear in mind that we are in the realm of partially shaded surfaces, and that to a certain extent the function of the coloured and the grey papers will be the same. We should also remember that the appearance of colour by candlelight is not the same as it is in daylight. Violet becomes grey, light-blue becomes greenish, dark-blue almost black, yellow borders on white, whilst white becomes yellow and yellow–red also becomes richer in its characteristic way, so that here also colours on the active side continue to be brighter and more powerful, whilst those on the passive side continue to be duller and more subdued.

In this experiment, therefore, the colours selected from the passive side have to be particularly bright and striking so that they can afford to lose something on account of this nocturnal operation. If small black, white, and grey images are now added to these coloured surfaces, their effect will be to bring about the properties already described. They will be distinct in so far as their brightness and darkness differ in degree from those of the colours. The case is the same when coloured images are added to black, white, and grey, as well as to coloured surfaces.

74.

We have varied the scope of these arrangements to the point of super-fluity in order to achieve certainty. For, this is undoubtedly the only means whereby the experimenter distinguishes himself from those who gape and gaze at random appearances as if they were uncon-nected occurrences. Newton, on the other hand, is constantly involved in getting his pupils to concentrate upon certain conditions, since other conditions would not bear out his point of view. The Newtonian presentation might, therefore, be likened to stage scenery, all the lines of which can only be seen as a proper and coherent whole from one single standpoint. Yet Newton and his pupils will not allow anyone to move a little to one side in order to peer into the wings. They keep the audience in their seats, and assure them that what they see is a real wall, closed and impassable.

75.

So far, we have reported the results of our precise observation of the matter, and it is evident on the one hand that the illusion was possible because Newton compared two coloured surfaces, a light one and a dark one — requiring that the dark one should behave the same way as the light one. He presents them to us as differing only in colour, paying no attention to the fact that they also differ in degrees of brightness. On the other hand, though, how he can say that black was visible on blue when it is not visible on red is quite incomprehensible to us.

76.

We have noted that, when one has found the position of the white screen which gives the most distinct image, the screen may be moved a little to and fro without appreciably disturbing its clarity. If it is moved a little too far forward or backward, however, the clarity does diminish and, if they are then compared, it is discovered that this is a consequence of the degree to which they contrast with the base, those which contrast the most retaining their distinctness the longest. Thus, white on black can still be seen quite distinctly when white on grey has become indistinct. Black can still be seen on red lead when it has disappeared on indigo blue, and under all conditions the same is true of the other colours employed. We have, however, been unable to find any evidence of a position in which a less contrasting image is more distinct than a more contrasting one, and we are therefore obliged to categorise Newton's assertion as an appearance which is the result of wishful thinking and ingrained prejudice, and seen only in the mind's eye. Since the apparatus is simple and the experiments easy to perform, others may be more fortunate in discovering something which might provide some justification for what was supposed to have been observed.

77.

(5) *Conclusion.* Now that we have demonstrated the worth of these assumptions, we regard ourselves as fully justified in proceeding straight to a denial of the conclusion. Indeed, we take this opportunity to draw the reader's attention to an important point which will often be raised from now on. It is that the Newtonian doctrine predicts far too much. For, if it were true, there could in fact be no dioptric telescope, and Newton did indeed argue on the basis of his theory that it was impossible to make any improvements to it; what is more, if this really was the situation, coloured objects in juxtaposition would appear to be thoroughly confused, even to the naked eye. Let us imagine a house standing in the full light of the sun; it has a red-tiled roof, it is painted yellow, it has green shutters, there are blue curtains behind the open windows, and a woman in a violet dress is coming out of the front door. Now, if we viewed the whole together with its parts from a certain position, from which we could take it all in at once, and the tiles were quite distinct, as soon as we turned our eyes upon the woman, we should not be able to make out at all clearly the cut and folds of her dress; we should have to move forwards, and if we then saw the woman clearly, the tiles would appear as if they were in a mist: so, in order for our eyes to take in wholly distinct images of the other parts, we would have to be constantly moving to and fro — if the alleged diverse refrangibility which was supposed to have been demonstrated in the second experiment were really taking place. It would be the same with all eye-glasses, simple or compound, and no less so with the *camera obscura*.

78.

Let us take a case directly related to the second Newtonian experiment, and ask our readers to call to mind that optical box in which brightly illuminated pictures of cities, castles, and squares are viewed through a lens and correspondingly enlarged, whilst at the same time also appearing extremely clear-cut and distinct. This can be said to be Newton's own experiment, only reproduced subjectively and with greater variety. If the Newtonian hypothesis were true, it would be impossible to see the juxtaposition of the light-blue sky, the light-green sea, the yellow- and blue-green trees, the yellow houses, the red-tiled roofs, the multicoloured cabs, liveries, and pedestrians, as being simultaneously distinct.

79.

Mention has to be made of several other odd consequences of the Newtonian doctrine. Consider those black figures placed on the differently coloured and comparably bright surfaces. The question arises as to whether or not the black image also has the right to determine its outline once it has passed through the lens. Two black images, one on a red base, the other on a blue one, are refracted to the same extent; for diverse refrangibility is not to be ascribed to the black. If both black images reach the white screen with the same distinctness, however, we would like to know how the red and blue base would behave in disputing their formerly clearly delineated boundaries. What is experienced is, therefore, in complete accordance with what we maintain, whereas the longer one occupies oneself with the Newtonian doctrine, experimentally or contemplatively, the more one becomes aware of its lack of truth and relevance.

80.

If one considers coloured figures on a coloured background, the so-called experiment and what is deduced from it becomes quite absurd. A red image could never appear on a blue background, and *vice versa*; for should the red boundary choose to become distinct, the blue one would not be so inclined; and, if it should finally get round to it, this would be inconvenient for the other one. If this were indeed the case with the elements of the doctrine of colours, Nature would have prepared for sight or the perception of visible appearances in a rather neater way.

81.

This is, therefore, the situation with respect to the two experiments which Newton regarded as so important that he took them to be the foundation pillars of his theory, and placed them at the beginning of a treatise, and the groundlessness of which scientists have been reluctant to examine for a 100 years — although the evidence and the objections we have produced have often been submitted, presented and worked out in print, as the *History* will make abundantly apparent.

Proposition II. Theorem II

The Light of the sun consists of Rays differently Refrangible

82.

Now that we have already become acquainted with coloured lights, which even the feeble light of the candle elicits from the surface of coloured bodies, now that we have already been introduced to what is derived or still to be derived, the author turns to the real source, namely, the sun as being a light we readily regard as being primordial.

83.

The light of the sun is said to consist of rays differing in refrangibility. Why, at this juncture should special mention be made of the sun? The light of the moon, of the stars, of any candle, or of any bright object on a dark background, is capable of displaying to us the phenomenon which is ascribed here to the sun in particular. The powerful action of the sun is certainly an advantage to the experimenter when it is used in those experiments we have allied objective; but it is in no respects an appearance basic enough to support a theory.

84.

That is why, in our treatment of dioptric experiments of the second class, we have given prime consideration to the subjective ones and thus made it abundantly clear that what is being discussed is neither light nor lights but an image and its boundaries; for the sun's image has no advantage at all over any other, not even over a light- or dark-grey one on a black base.

85.

According to the Newtonian doctrine, however, colours really are hidden within light and have to be developed out of it. Even the title of the work makes this point, for here, too, our attention is called to the "colours of light", as the Newtonians have always been in the habit of calling them. It is not surprising, therefore, that this phrase should also be introduced here. The author devotes eight experiments to attempt to prove this fundamental premise of his chromatic theory; let us now investigate his procedure by examining experiments three to ten in the order in which they are presented.

Third experiment

86.

We will not examine this part of the author's text word by word; for this is the well-known experiment where the sun's image is allowed to pass through a small opening in a window shutter into a dark chamber, and then through a horizontally oriented prism, the refracting angle of which points downwards; the image which is thrown on to the opposite wall is refracted upwards, and is no longer colourless but elongated and coloured.[8]

87.

All those who are familiar with our general treatment of dioptric colours of the second class and, more particularly, with the detailed account of the objective experiments of Chapters 20 to 24, will know precisely what this phenomenon involves. We would call particular attention to the second, fifth, and sixth of our plates, for they demonstrate clearly that it is by no means a fixed appearance which

---

[8] See Figure 13 in the Appendix.

emerges from the prism, for if the screen on to which it is supposed to be projected is shifted backwards and forwards, each move causes it to display new relationships. As soon as anyone realises this, there is no need to enter into any further dispute with him, either concerning this third experiment or the Newtonian theory as a whole; for, although the greatest importance is ascribed to this experiment by both the master and his pupils, Plate V[9] of our work shows clearly that they totally fail to present it correctly.

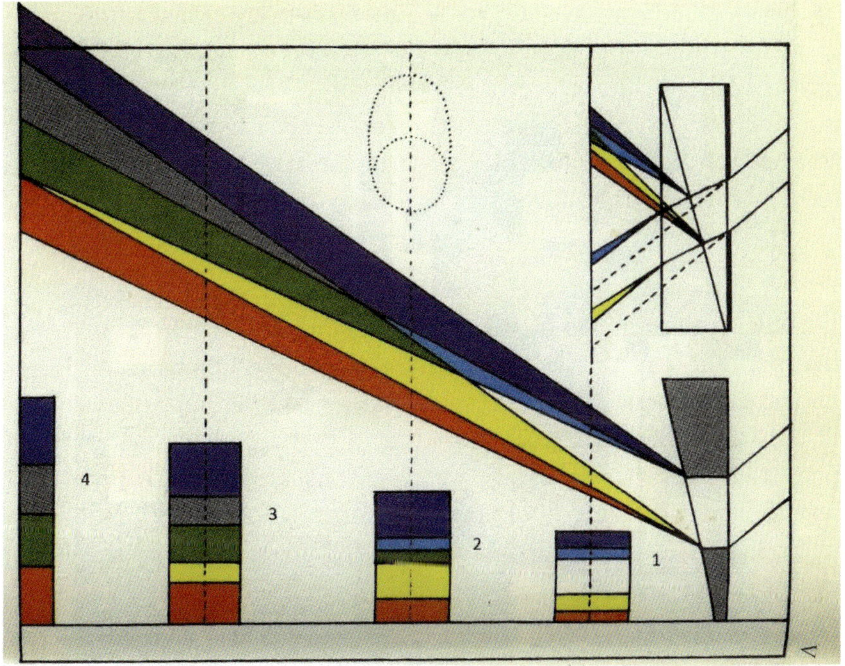

Plate V

88.

They assert, quite contrary to the truth of the matter, that the phenomenon which emerges from the prism is fixed, that the colours one sees in the elongated image are always in the same sequence and

---

[9] Goethe provided 14 explanatory plates for *Zur Farbenlehre*, Plates I to VI relate to *Outline* and Plates VII to XIV to the present *Exposure*.

proportion. As the screen is moved further away, the image lengthens in this sequence and proportion until it is approximately five times as long as it is broad, in which condition they finally decide to hold it fast. When they have fixed it in this position, observed, measured, and manipulated it in various ways, they draw the following conclusion: if in the round image, which they refer to as the image of a ray, all parts were equally refrangible, then they must all arrive at the same place after the refraction, and the image would therefore appear the same as before. But the image is now elongated; consequently, some parts of the so-called ray have lagged behind whilst others have hurried on, and they must therefore possess a diverse determinability during refraction, and thus a diverse refrangibility. What is more, this image is not white but multicoloured, and can be seen as a series of colours, from which they conclude that each of the rays which is supposed to have its own refrangibility must also have its own particular colour.

89.

We have at present, no further comments to make, apart from noting that the whole of this argument is founded upon an incorrectly interpreted experiment, which does not display itself naturally as it appears in the book; the main point that has to be made is that what is displayed by the prismatic image which emerges from the prism is in no way a fixed series of colours but a coloured appearance separated by white light. Since Newton and his pupils completely failed to develop this phenomenon as they should have done, its true nature necessarily eluded them and error was piled upon error. It is recommended that the reader pays particular attention to the following paragraphs.

90.

Newton, after having painstakingly measured the appearance and observed many of its accompanying properties, though not the right ones, continues:

> *The different Magnitude of the hole in the Window shut, and different powers of the Prisms[10] which the Rays pass through (and different*

---

[10] Stated in the singular in *Opticks.*

inclinations of the prism to the Horizon)[11] *make no sensible changes in the length of the Image.*

91.

Both of these assertions are quite untrue, for it is precisely the size of the image and the refracting angle[12] of the prism used which play the major part in determining the extent to which its length is greater than its width. We shall devote a figure in our plates[13] to the first of these two effects, and will now proceed to indicate what is required for a more accurate insight into this relationship.

92.

Our attentive readers will be aware that when a light image is displaced, the yellow–red edge and yellow border extend into it, whereas the blue-edged violet border extends beyond it. The yellow border can never extend further than the blue edge opposite — with which it combines to form green; and this is in fact the limit of the inner edge. The violet border, on the other hand, continues to pursue its course, steadily widening as it does so. Thus, if a small hole is taken, its illuminated image can be displaced until its length appears to be five times greater than its width; but this is in no way a typical elongation for larger images under the same circumstances. Take a piece of cardboard or sheet metal which has several holes of various sizes, the tops of which are ranged in a horizontal line. Slide it in front of a water prism and allow the sunlight to play upon these various holes so that the images refracted by the prism project on to the wall, but in such a way that, since the tops of the holes are in a

---

[11] Omitted by Goethe.

[12] Newton's wording was "and different thickness of the prism...", meaning that, since the refraction of the ray occurs at the surfaces of the prism, the thickness through which the ray passes, whether it be near the apex or base, plays no role. Both "thickness" and "power" can be translated as *Stärke*, Goethe, using Klugel's German translation, would seem to have assumed the latter meaning (i.e., "power") and consequently misunderstood Newton's statement. The word has therefore been back translated to "power" instead of Newton's own "thickness" to register this point.

[13] Goethe did not in fact do this.

horizontal line it has a different relationship to this border, its width not being contained so many times in its length as is the case with the small image. It is also very easy to perform this experiment subjectively by fixing a series of different sized white discs on to a black screen. In this case, however, since the refracting angle is usually turned downwards, it is the bases of the discs which have to form the horizontal line.

93.

That the power of the prism, that is to say, the size of its refracting angle, must affect the relationship between the length and the width of the image, will moreover be evident to anyone aware of what has been noted in paragraphs 210 and 324, especially with respect to the third condition, and to what has been demonstrated in more detail in other parts of the treatise, namely, that a greater displacement of the image is a primary condition for stronger colouring. Now, since a prism with a greater refracting angle displaces an image more than one with a smaller one, different coloured appearances will occur under the same conditions. We shall now, without further ado, make it perfectly clear to the reader what the significance of this experiment is, and what it is capable of proving.

Fourth experiment

94.

If the observer now looks through the prism at the image of the sun projected into it or at the opening which is only illuminated by the light of the sky, he will be converting the previous objective experiment into a subjective one. If the procedure were of any help to us there could be no objection to it, but the subjective image can no more be traced back to its origins than the objective one could. The observer only sees the elongated and continuously coloured image, the violet part again being the most extended.

95.

Unfortunately the author also conceals an important point from us in this case, namely, that the appearance is the exact reverse of the one we have just observed on the wall. If one realises this, one naturally asks what would happen if the eye were positioned in the place of the screen. Would the colours then be seen as they are on the screen, or in the reverse order? And what, then, is the actual overall relationship?

96.

This question was already raised in Newton's time, and there were individuals who opposed him in maintaining that when the eye looks from the screen to the source, it sees precisely the opposite colour to that projected on to the screen or into the eye positioned in the place of the screen. Typically enough, Newton gave no consideration to this objection — he simply rejected it.

97.

The true relationship is, however, that the two images have nothing in common. They are quite different, one being moved upwards and the other downwards and, in accordance with the regular law, they are therefore coloured differently.

98.

There are numerous ways of convincing oneself of the co-existence of these two different images, the objective one being coloured upwards and the subjective one downwards. The following experiment is, however, certainly the easiest and the most impressive. By means of a two or three inch opening in the window shutter, the sun's image is allowed to pass through a large water prism on to some thin white paper held taut in a frame so that the two coloured fringes of the image are still separate — there being as yet no green, the centre still being white. On observing this image behind the frame, one sees the blue and violet quite clearly at the top and the yellow–red and yellow at the bottom. If one now looks through the prism from the side of the frame, however, the image of the opening in the shutter, which is displaced downwards, will be seen to be coloured in the inverse order.

By the following means, one is able to observe both images not only superimposed on each other but also juxtaposed. Spirit of soap is dropped into the water in the prism until the turbidity of the water is such that, although the image on the picture frame is still distinct, the sunlight is so moderated that the eye is not dazzled by it. Positioning oneself behind the frame, one makes a small opening in the paper where the refracted and coloured image has formed and, on looking through it, one will see what one saw previously, namely, the sun's image displaced downwards. If the hole made in the paper is big enough, one can then step back a little and at one and the same time see both the objective, translucent image which is coloured upwards and the subjective image which is displayed in the eye. What is more, by moving the paper up and down a little, one can bring the similar and dissimilar fringes of the two appearances together to whatever extent is required; and in that one convinces oneself of the co-existence of both appearances, one also convinces oneself of their constant mobility and modulating effectiveness. One is reminded of the very remarkable experiment described in 0. 350–354, which we should be thoroughly familiar with since it will often be referred back to in what follows.

Fifth experiment

99.

Newton's conception of this experiment is also clouded by prejudice. He has no clear idea of what he is looking at or of what can be deduced from the experiment. The appearance is, however, very useful to him since it lends credence to his demonstrations, and he constantly refers back to it. What happens is that the spectrum, that is to say, the elongated image of the sun which was formed by a horizontal prism in the third experiment is intercepted by a vertical prism and thus refracted sideways. It appears precisely as it did before, except that it is bent forward to some extent, with the violet part leading.[14]

100.

Newton draws the following conclusions from this:

If the cause of the elongation of the image lay in the sun's rays having been dispersed, split, and dilated in refraction, such an effect would be reproduced by a second refraction, and the oblong image, intercepted lengthwise by a second prism, parallel to the axis of the same, would once again, as on the previous occasion, be spread sideways and dilated. But this does not happen, for the image emerges unchanged as an oblong, and simply becomes somewhat oblique; from which it appears that the cause of the appearance lies in the fact that light has a certain property which, having once manifested itself as so many coloured lights, is no longer capable of influence, with the result that the phenomenon is unchanged, except that it becomes more oblique after a second refraction, but in a manner entirely in accordance with Nature, for here too the more refrangible violet rays are those which advance, thus again letting it be seen how they differ from the others.[15]

101.

The error Newton commits here is the one we have already drawn attention to and which he commits throughout the entire work, namely that of regarding the prismatic image as being something fixed and immutable, whereas it is in fact merely in a state of constant modulation and change. Whoever fully grasps the difference here has understood the sum total of the dispute. They will not only see the point of our objections and agree with our having raised them, but will also be able to develop them for themselves. We have already prepared the way for obtaining an insight into the relationship exhibited by this phenomenon in the first part of this work (O. 203–207),

---

[14] See Figure 14 in the Appendix.

[15] Goethe has carried out a free translation on this occasion, but the meaning is retained.

and our second plate will also be found helpful in this respect. Prisms with small refracting angles of about 15° should be used, so that the modulation of the image may be clearly observed. If the image is subjectively displaced by a prism so that it appears to be raised upwards, it will be coloured in this direction. If two prisms are now superimposed crosswise the image must, in accordance with the general law, be displaced diagonally, and thus be coloured in this direction. In all these cases the image is a modulating and developing formation, for the edges and borders only develop along the line of displacement. Newton's inclined image, however, is in no way the one that was collected originally, making a bow after the second refraction, but an entirely new one which is coloured in the requisite direction. If one turns back to the paragraphs and tables in our work which have already been referred to, one will benefit to the extent of being fully convinced that this is the case.

Thus prepared, one will discover one false inference after another, and if one then goes on to examine Newton's own so-called illustration of this experiment, and the figures and descriptions he devotes to it, one will be unable to avoid the conviction that this experiment lends no credence at all to the proposition.

102.

Whilst leaving our readers to undertake these investigations on their own let us now call their attention to the peculiar procedures the author proposes to adopt.

103.

In the fifth experiment the prismatic image appears to be not only lowered but also elongated. This can readily be accounted for by our theory, for we need a second prism in order to make the image appear in the diagonal, and this is tantamount to saying that it is as if the appearance had been created by a double prism. Now, since one of the main requirements for the extension of the appearance of colour is the increased effectiveness of the medium (O. 20), then the increase in the length of this image must be proportional to the strength of the prisms used. This explanation should be borne in mind when attempting to show how cunning Newton is when he goes about gaining his end.

Our readers will be aware how the coloured image which is brought about by refraction may be further elongated, for we have covered the conditions for this in some detail. They will be no less aware that, since there is a constant broadening of the coloured edges and borders as the image is elongated, so that the opposite ends of the image are constantly converging, this elongation is also accompanied by an increased merging of the image's contrasting elements. We are quite happy to state this in a straightforward manner, for such is the nature of the matter.

Newton, however, with his fabricated parody of Nature, has to busy himself in all kinds of ways, heaping up experiment upon experiment, fiction upon fiction, in order to bedazzle where he cannot persuade.

And yet, at this juncture he concerns himself with the proof of his second proposition, namely, that sunlight consists of rays differently refrangible. Since these different light rays and lights are supposed to be the integrating constituents of sunlight, our author is well aware that it can and must be required that these various components should also be seen to be separated out and clearly distinguished, one from another.

The ordinary spectrum, the phenomenon examined in the third experiment, has already been accounted for as a demonstration and manifestation of the forcing asunder of the various lights in sunlight, the drawing apart of the various coloured images in the image of the sun. Yet, what we have here is still far removed from separation. To separate out, carve up, tear apart a continuous series of intermingling colours which are welling forth from one another as it were, is a difficult business; and yet the problem is raised by Newton in his fourth proposition: the heterogeneous rays of compound light are to be separated from one another. Since he has set himself an impossible task, he certainly has to make an early start if he is to succeed in the step-by-step business of leading the unsuspecting pupil astray. It is worth noting how he goes about it.

104.

But that the meaning of this Experiment may more clearly appear, it is to be considered that the Rays which are equally refrangible do fall upon a Circle answering to the sun's Disque. For this was proved in the third Experiment.

105.

Had it been proved, no objection could have been raised; for it would have been natural if the constituents streaming from the sun were variously refrangible, even if they came from one and the same solar disc, some of them would have lagged behind after refraction whilst others would have gone ahead. We are already aware, however, that this is not the case. Let us read on.

106.

By a circle I understand not here a perfect geometrical Circle, but any orbicular Figure whose length is equal to its breadth, and which, as to Sense, may seem circular.

107.

Such special pleading and persistent appealing, as it might be called, goes on throughout the whole of Newtonian optics. For he first says something and establishes it; then, since it is only in apparent accordance with experience, he proceeds to qualify his proposition until he has completely nullified it. Opponents have often called attention to this procedural mannerism, yet the school has been neither capable of taking note of it nor prepared to answer for it. For further insight into the question we now turn to Figures 4, 5, 6, and 7 of our seventh plate.

The fourth figure shows the spectrum as Newton and his followers in that captious way of theirs often present it: as an oblong figure bounded by two parallel lines, rounded at the top and bottom, but no mention is made of any colour. It is Figure 5, however, which relates to the present discussion.

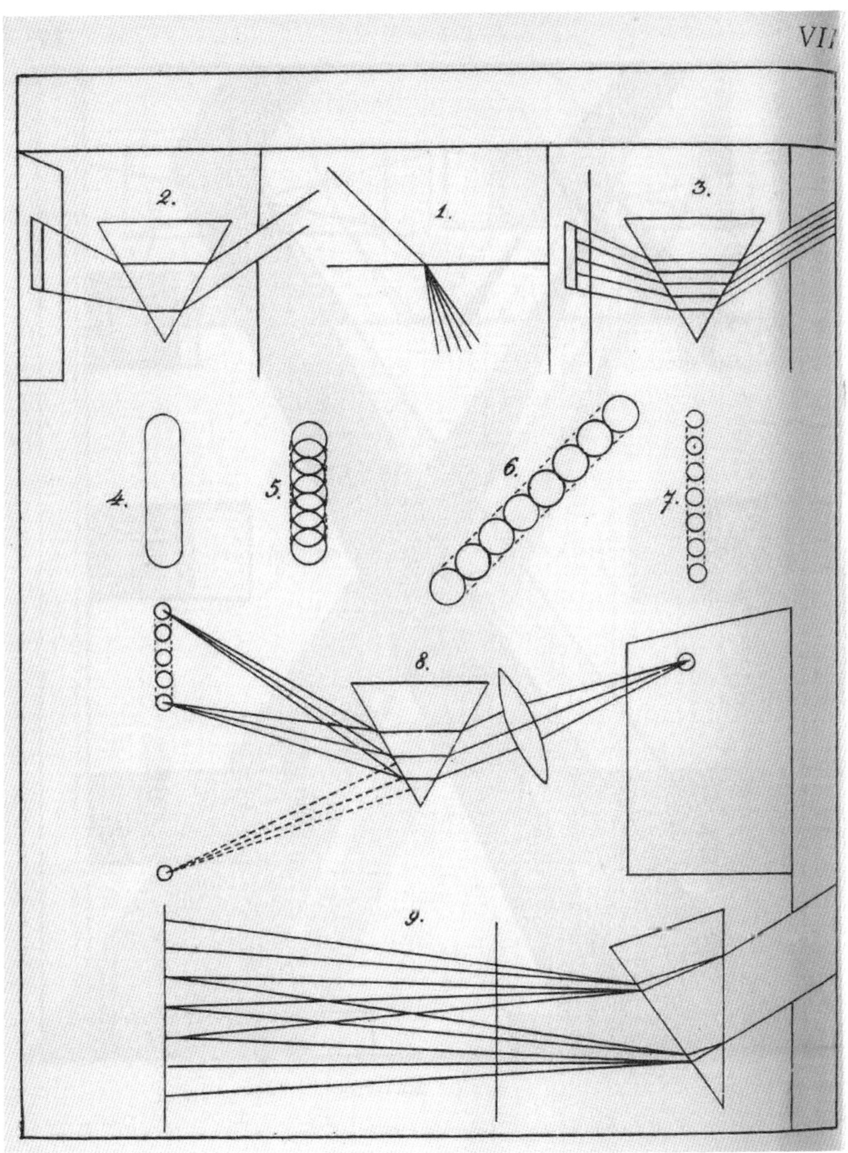

Plate VII

108.

Let therefore the upper Circle represent that in which all the most refran-
gible Rays propagated from the whole Disque of the sun, would

illuminate and paint upon the opposite Wall if they were alone; the lower Circle represent that in which all the least refrangible Rays would in like manner illuminate and paint if they were alone; the intermediate Circles represent those in which so many intermediate sorts of Rays would successively paint upon the Wall if they were singly propagated from the sun in successive order, the rest being always intercepted, and conceive that there are other intermediate Circles without Number, which innumerable other intermediate sorts of Rays would successively paint upon the Wall if the sun should successively emit every sort apart. And seeing the sun emits all these sorts at once, they must all together illuminate and paint innumerable equal Circles, of all which, being according to their degrees of Refrangibility placed in order in a continual Series, that oblong Spectrum is composed which I described in the third Experiment.

109.

This is a hypothetical representation, the hieroglyph of a conviction, and in no respect an image of Nature; if the reader is interested he can easily find out for himself how its author employs it in order to explain the bowing performed by his spectrum. Our immediate task is merely to make clear how untenable the concept is. Intermeshing discs are simply not present; the illusion that they can only arise when the refracted image is round, for then the boundaries of the coloured image, as an accessory image[16] also appear to be round, the reason being that in prismatic experiments the progress of the various components of the coloured image always takes place in parallel lines, which at any point in time intersect the line of advance at right angles. In order to make this clear, we have assumed in our fifth and sixth plates that it is a rectangular image which is displaced, for one can then form a clear picture of the parallel advance of the various series of colours. We are therefore obliged to emphasise once again that this is a matter neither of five or seven intermeshing discs nor of discs innumerable, but of a red edge forming at the boundaries of the image

---

[16] *Nebenbild.* This Goethean word arises from his notion that, since double images often appear in reflections from thick glass plates, then by analogy they should also appear on refraction; see (O. 220–242).

which merges into yellow, of a blue edge which merges into violet. If, because of the narrowness of the image or the power of the refraction, the yellow border reaches the blue edge across the white image, green is formed; if the violet border reaches the yellow–red edge across the black image, purple is formed. One cannot only see this with one's own eyes, one is almost tempted to say that it is corporeal.

110.

Since Newton is well aware of the need to separate his differently refrangible rays even more widely, he is not content merely to wrench them apart whilst still allowing their discs to intermesh. In the second figure he shows them as being isolated from each other; but, by retaining the boundary lines between them, he presents them as being separated and yet connected. Both diagrams appear in Figure 15 on Newton's third plate. The sixth figure on our seventh plate illustrates this ostensible pulling apart of the discs. We will return to this matter subsequently.

111.

We now wish to draw the research student's attention to the manner in which the author makes the transition to the next experiment. One prism was placed above another and an image of the sun allowed to pass through each simultaneously and then intercepted by a vertical prism so that each was bent sideways.[17] Probably because the vertical prism was not long enough to catch two fully formed spectra he moved it nearer to the first two prisms, discovering what we have long known and understood, namely, that even after refraction the two images are round and rather colourless. He is not put out by this at all, however, for instead of realising and admitting that he has been interpreting the matter quite erroneously, he remains totally unperturbed, simply observing that:

112.

This happens whether the third Prism be placed immediately after the two first, or at a great distance from them, so that the Light refracted in

---

[17] See Figure 17 in the Appendix.

the two first prisms be either white and circular, or coloured and oblong when it falls on the third.

113.

Suddenly, therefore, we have a coloured image which has passed through a prism and been refracted and yet is still white and round, whereas up until now it has only been presented to us as oblong, pulled apart and completely coloured. How is it that white has now managed to slip in through the back door? What is the explanation? In light of the foregoing discussion how is it possible? This is but one example of the extremely pernicious kind of special pleading so characteristic of Newtonian optics. By means of an incidental observation we are suddenly confronted with light which is refracted and yet white, compound and yet not resolved into its element's by means of refraction. No one notices that the appearance of this white discredits the whole of the foregoing exposition, that if one is to get to the heart of the matter one has to go elsewhere, begin all over again in a quite different way. The author, however, who has turned so suddenly down this new road, continues calmly on his way; in addition to his fully formed spectra with a green centre he now has a nascent and as yet uncoloured spectra with a white centre, as well as an oblong and a round spectra with which he operates alternately as the fancy takes him; and the world which has now been idolising his doctrine for a hundred years not only remains unaware of the juggling that is going on, but even hounds down and manhandles those who want to expose it for what it is.

The observation is dragged in at this juncture because the author is in dire need of this white centre, which suddenly pops up at this point in order that he may proceed with his hocus pocus.

Sixth experiment

114.

Until now we have been circumspect and guarded, and have reacted briskly — sharply even, whenever Newton has confronted us with experiments which appeared to have been selected with the express purpose of deceiving us and hustling us into ill-considered acquiescence. Now we have to be particularly careful in this respect, for we have to deal with the experiment which first convinced Newton that his explanation was the correct one, and which does indeed appear to carry more weight than all the others. This is the so-called *experimentum crucis*, by means of which the investigator stretched Nature upon the rack in order to make her confess what he had already fixed his mind upon.

Nature is not lacking in resolution and high mindedness, however, and sticks to the truth even under torture. If what is recorded does not confirm this, the reason is that the inquisitor has not heard correctly, that what the scribe has recorded is something other than the truth. A falsified pronouncement of this kind may be accepted for a while, but sooner or later someone will choose to take the part of wronged innocence; as we have, in thus donning our armour for knightly combat in the service of our damsel. We shall now begin by examining the way in which Newton approaches the matter.

115.

In the middle of two thin Boards I made round holes a third of an Inch in diameter, and in the Window shut a much broader hole being made to let into my darkened Chamber a large Beam of the sun's Light; I placed a Prism behind the Shut in that beam to refract it towards the opposite Wall, and close behind the Prism I fixed one of the Boards, in such a manner that the middle of the refracted Light might pass through the hole made in it, and the rest be intercepted by the Board.[18]

116.

Newton sallies forth in his usual way. He specifies conditions, but not the reasons for them. Why, suddenly, is there a large hole in the shutter, the prism presumably also being large, although he makes no mention of this? The size of the hole gives rise to a large image, and

---

[18] See Figure 18 in the Appendix.

after refraction a large image which is cast on to a board placed just behind the prism also has a white centre. So, here we have the white centre which was slipped in surreptitiously at the end of the previous experiment (112). He operates within this white centre: but why does he not mention that it is white. Why does he leave us to stumble upon this important particular? The reason is, of course, that his entire theory collapses as soon as it has to be accounted for.

117.

Then at the distance of 12 Feet from the first Board I fixed the other Board in such a manner that the middle of the refracted Light which came through the hole in the first Board, and fell upon the opposite Wall, might pass through the hole in this other Board, and the rest being inter-cepted by the Board might paint upon it the coloured Spectrum of the sun.

118.

Once again, therefore, we have a centre of refracted light and, as is clear from the last statement, it is a green centre; for the rest should certainly display the coloured image. We are presented with two centres, a colourless and a green one, without the slightest intimation being given of the difference between them — an important differ-ence, moreover, on which everything depends. It would take a lot to open the eyes of anyone who, even in a case such as this, fails to see anything odd in the Newtonian procedure. But we will not pursue the matter, for, in spite of it being described so precisely, the set-up is quite superfluous, as we shall see when we come to examine the illus-tration of this experiment. Let us now proceed directly to this exami-nation, skipping some of the text, the content of which is in any case repeated in what follows. The object of our twelfth plate[19] is to create a better understanding of the matter, and our readers should there-fore have it to hand. Among the figures it contains they will find one taken from Newton which involves misrepresentation and a correct

---

[19] Never published by Goethe. There is a Plate 12; but it clearly does not apply to this experiment.

one which presents the experiment in its true light. In order to facilitate comparison, both have been lettered in the same way.

119.

Let F be the wide hole in the Window shut, through which the sun shines upon the first Prism ABC, and let the refracted Light fall upon the middle of the Board DE and the middle part of that Light upon the hole G made in the middle part of that Board. Let this trajected part of that Light fall again upon the middle of the second Board <u>de</u>, and there paint such an oblong coloured Image of the sun as was described in the third experiment.

120.

On the first occasion, therefore, the middle part is white; this has already been mentioned above, although once again it is passed over in silence by the author. We now ask how it is that the white part which falls on the board DE, by passing through the hole G, gives rise to a fully coloured image on the second board *de*? The answer must be that this is due to the attenuation undergone by the light image after refraction within the small hole G. The unavoidable implication of this would be, however, that an attenuation, a limitation, is necessary for the prismatic appearance of colour; and yet this conclusion is stubbornly resisted in the second part of the book. Newton has to remain silent about these relationships, these necessary and indispensable conditions; if the reader or pupil is to continue to believe in what is going on, he has to be kept in the dark. Our figure gives the clearest possible illustration of the fact of the matter, however, making it perfectly clear that the coloured borders arise through the effect not only of the first but also of the second hole and that, since the second hole is small enough, these borders very soon overlap as they spread out, and so display the fully coloured image. After this contrivance, Newton drives on to his goal.

121.

By turning the Prism ABC slowly to and fro about its Axis, this Image will be made to move up and down the Board <u>de</u>, and by this means all its *parts from one end to the other may be made to pass successively through the hole g which is made in the middle of that Board. In the meanwhile another Prism abc. is to be fixed next after that hole g, to refract the trajected Light a second time. And these things being thus ordered, I marked the places M and N of the opposite Wall upon which the refracted Light fell, and found that whilst the two Boards and second Prism remained unmoved, those places by turning the first Prism about its Axis were changed perpetually. For when the lower part of the image which fell upon the second Board de. was cast through the hole g, it went to a lower place M on the Wall and when the higher part of that Light was cast through the same hole g. it went to a higher place N on the Wall, and when any intermediate part of the Light was cast through that hole, it went to some place on the Wall between M and N. The unchanged Position of the holes in the Boards, made the Incidence of the Rays upon the second Prism to be the same in all cases. And yet in that common Incidence some of the Rays were more refracted, and others less. And those were more refracted in this Prism, which by a greater Refraction in the first Prism were more turned out of the way, and therefore for their Constancy of being more refracted are deservedly called more refrangible.*

122.

Newton himself gives the reason for his having used two perforated boards in this experiment. He wants to show that the angle of incidence of the rays falling on the second prism is unaltered by any movement of the first; but he overlooks, or is concealing from us, what we have already noted, namely, that the coloured image first arises behind the hole in the first board, and that the origin of its various parts, in that they pass through the hole in the second board, can still be attributed to a variation of incidence upon the second prism.

123.

Although many of Newton's earlier opponents regarded the effect of incidence in the experiments as important, we are not in agreement with them: we simply mention this particular in order to make the point that in this, as in other experiments, certain pernickety arrangements are by no means essential. In the present case the two boards are extremely inconvenient; they produce a reduced and weakened image, with which it is impossible to work clearly and well. And, although the result does eventually appear, it often remains indistinct because of the cumbersome apparatus, and it is difficult for the experimenter to set up the whole contrivance with the necessary precision.

124.

We do not deny the appearance, therefore, but simply attempt to come by it in another way, to relate it to what the following experiment will show us, as well as to what our earlier experiences have already established; we ask our readers to be particularly attentive as we do so, for we are now nearing the axis upon which the whole controversy turns, the point on which the Newtonian doctrine either stands or falls.

125.

Our readers are already sufficiently acquainted with the various conditions under which the prismatic image elongates itself in both subjective and objective cases (O. 210, 324). By and large, they may be categorised under the principle condition of it being increasingly displaced from its position.

126.

If the coloured image, with all its parts, which has passed through the first prism and appeared on the board is now intercepted by and passes through a second prism so that it is again displaced from its path; it is not only elevated but also elongated. What happens when the image is elongated? The spread of the various colours increases, the colours separate out further according to certain proportions.

127.

On displacement of a bright image, the yellow–red edge in no way advances to the extent that the violet border does; the latter moves away from the former. The entire spectrum formed by the first prism might, when measured, be found to be three inches long and the centre of the yellow–red to be about two inches from the violet. If this spectrum is then refracted through the second prism so that it lengthens to about nine inches, the distance between the centres of the yellow–red and the violet will also be much greater than before.

128.

What holds for the whole image also holds for its parts. If the coloured image brought about by the first prism is intercepted by a perforated board and the resulting appearance, which consists of various isolated coloured images, is allowed to fall on to a white board, these separate images, which are indeed simply a whole spectrum broken into parts, take up the positions they occupied previously within the sequence of the whole.

129.

If this interrupted image is now intercepted just behind the perforated board by a prism and refracted a second time, the separate images, since they shift further upwards, will spread to varying extents — particularly violet, in that the progressive border, will distance itself proportionately more than the others. All that one has here, however, is the regular elongation of the whole image of which, finally, only parts can be seen.

130.

This is not so obvious from Newton's arrangement; but cause and effect remain the same, even if by rotating the first prism he sends the images through the second prism separately; but they are still parts of the whole coloured image and remain true to their nature.

131.

There is, thus, no diverse refrangibility here, simply a repeated refraction, an increased elongation, nothing more and nothing less.

132.

Any vestige of doubt can be removed by performing the experiments with a dark image, in this case the yellow border advances and the blue edge holds back. All that previously applied to the violet part now applies to the yellow, and what was said about the yellow–red is now true of the blue. Whosoever has seen this with his own eyes and given it proper consideration will find the supposed significance of this key experiment vanish away like mist. Whatever may appear to be necessary in order to make this even clearer will now be brought up in the course of explaining our twelfth plate, and we shall also give a precise description of the apparatus required for this experiment.

133.

The only further comment we have to make here concerns the captious way Newton presents the matter (121) when he says that, as a result of the second refraction, the small red image went to the lower part of the wall[20] the violet image to the upper part. (The word in English is "went", in Latin it is *pergebat*.) For this is by no means what happens. After the second refraction, both the yellow–red and the violet part shift upwards, although the latter moves away from the former to the extent that the image would have grown had it been refracted as a whole and not in parts.

134.

Now, since this experiment is not backed up by any other and proves nothing, it does not even need to be derived from anything or explained, being nothing more than a perfectly well-known phenomenon; in fact, the matter can be settled easily by referring to our explanation in the *Outline*, one might well reproach us for not addressing ourselves to the matter immediately by dealing initially

---

[20] Newton in fact wrote "it went to a lower place...on the wall" not "...the lower part..." implying that red only went to a lower part of the wall relative to violet. Goethe has altered the meaning.

with this so-called key experiment and demonstrating the inadmissibility of the argumentation based upon it, instead of entering upon the roundabout business of following the Newtonian deduction step-by-step and accompanying the author down all his blind alleys. Our answer is that there is no point, when attempting to destroy a deep-seated prejudice, in simply making known one's own central insight. Pointing out that the house is dilapidated and uninhabitable is not enough, for it can still be shored up, and furnished, if need be; nor is it enough simply to gain access and demolish the place: the rubble has to be carted away, the site cleared and levelled. Only then is there a good chance that certain interested parties will set about erecting a new workmanlike structure.

135.

It is with this in mind that we now proceed to diversify[21] the experiment. The following arrangement is useful in that it illustrates the phenomenon in question particularly clearly: two identical prisms are positioned close to each other and behind them is placed a board in which two horizontally aligned small holes are made. The yellow part of the image formed by one prism is allowed to play upon one hole, and the violet part formed by the other prism to fall upon the other hole. If the two differently coloured images are intercepted by a white board placed behind the prisms, the images will appear on it horizontally juxtaposed. If a prism which is large and long enough to intercept both small images is now placed immediately behind the perforated board so that both images are refracted a second time, they will once again be displayed upon the white board. Although the upward displacement of both is apparent, it is unequal, the violet moving up much higher than the yellow–red, as is to be expected from the explanation already given. We recommend this experiment to all remaining Newtonians who wish not only to intrigue their pupils but also to fortify their faith. Anyone who has patiently followed our exposition, however, will know that what happened here to

---

[21] *Vermännigfaltigen*: a key Goethean word, being the essential principle of his method (after Bacon).

the separate parts is simply what would happen to the image as a whole if one were placed lower than the other and both underwent a second refraction. The latter experiment can easily be set-up by using a water prism.

136.

Incidentally, we find it necessary to mention yet another point, which will be discussed in full when dealing with the next experiment, it arises in this one, too, but not with the same significance. It is that the images brought about by the objective action of the prism ought always to be regarded as modulating and mobile, and we have regarded them as such throughout; one cannot operate with them without altering them. One can, however, adopt the Newtonian approach of treating them as fixed and operating with them as such, but only under very special circumstances and only for a moment.

137.

If we now regard the isolated images which have passed through the perforated board as being fixed, operate with them and displace them by means of a second refraction, what we have demonstrated with regard to the displacement of coloured images in general is bound to take place; that is to say, they will acquire the usual edges and borders, although these will be either heightened or toned down by the colour of the image. From the yellow–red edge as it extends inwards we obtain an isolated yellow–red image; on its lower boundary its colour is strengthened by a new edge of the same kind, the barely emergent yellow fading away and, because of the opposition, no blue and hence no violet being able to arise on the opposite side. Consequently, the yellow–red remains forced back into itself, as it were — its appearance being less extended and significant than it might be expected to be. The violet image, on the other hand, is part of the outward-spreading violet border of the whole image. Although it may be a little curtailed on its lower boundary, it has complete freedom of advance at the top. This, taken with the observations made above, makes it likely that what takes place here is also a further displacement of the violet. We attach no great importance to this, however, and simply mention it in connection with such a complicated matter in order that each ancillary condition may be borne in mind; thus, in order to satisfy oneself as to the origin of these new edges, one has only to pass the yellow part of an image through a hole in a board and then refract it a second time.

Seventh experiment

138.

In this experiment the author places two prisms adjacent to each other and allows their two spectra to be cast into a dark chamber. The red colour of one spectrum and the violet colour of the other fall adjacently on to a narrow horizontal strip of paper. He then views this doubly prismatically coloured strip through a third prism and finds that the paper appears to be torn in two. The blue colour of the strip has, of course, shifted much further down than the red; the assumption is that the observer is looking through a prism, the refracting angle of which is pointing downwards.

139.

This is evidently meant to be a repetition of the first experiment, in which use was made of corporeal colours; here, however, we have apparently coloured surfaces which have acquired an appearance of being transmissive. The present case, the present arrangement is, therefore, very different from the former one and, since we do not deny the phenomenon, we shall attempt to demonstrate the hollowness of the Newtonian interpretation of it by reassessing it in various ways, by deriving it from our sources.

140.

We can continue to employ the arrangement already described (135), in which use was made of two adjacent prisms. The two small red and violet images are allowed to fall adjacently on to the white screen, being exactly aligned horizontally. A horizontal prism with its refracting angle pointing downwards is now raised to the eye and the screen observed through it; although it will be displaced in the usual way, one will also observe the appearance of a significant variation; namely, that the displacement of the red image corresponds with that of the screen; it always exactly retains the same position on the screen. The violet image does not behave in this way; this image alters its position, moves further down, and no longer forms a horizontal line with its red counterpart.

141.

Had the Newtonians still been able to confine the theory of colours to the dark room, to keep their followers in leading strings and deny them the opportunity of making any observations of their own, we would have made a particular point of recommending this experiment to them, for it involves something which is both surprising and impressive. Our task is to clarify the relationship of the whole, however, and to demonstrate in the present experiment that which was established in the last one.

142.

Here, for the first time, Newton combines objective and subjective experiments. He would have been well advised to begin by performing and interpreting the key experiment (O. 350–356) but in his unmethodical way he makes no mention of it until much later, in a context in which the phenomenon simply creates the possibility of fresh confusion, doing nothing to establish a true understanding of the matter. We shall assume that everyone has witnessed the experiment, that all who are interested are so equipped that they are able to repeat it whenever there is sunshine available.

143.

In this case, a prism casts the elongated coloured image upwards on to the wall; an identical prism is then taken, with its refracting angle pointing downwards, held in front of the eye and the screen approached. This does not appear to cause much of a change to the image; as one retreats, however, the image not only moves further down, but also withdraws further into itself in such a way that its violet border becomes progressively shorter. Eventually, the centre appears to be white and the colour to be only on the boundaries of the image. If the distance at which the observer is standing is precisely the same as that between the screen and the prism which gave rise to the coloured image, this image will appear subjectively to him to be colourless. If he moves further back, it becomes coloured downwards in the reverse order. When one is twice as far from the wall as the first prism, one sees both the coloured image spreading upwards with the naked eye and, through the second prism, the equally strongly coloured reverse image which is spreading downwards. From this, it is clear once again that the primary image on the wall is in no way a fixed entity, immutable both as a whole and in its parts but, although it is found to be defined, it is also found to be definable and, indeed, definable to a reversed condition.[22]

144.

Now, what is true of the whole image is also true of its parts. Let this entire image, before it reaches the screen, be intercepted by a perforated board in such a way that one is able to observe simultaneously both the whole image on this interposed board and also the isolated differently coloured images on the main screen. The previous experiment is then repeated. One moves up closely to the main screen and observes the vertical series of isolated coloured images through the horizontal prism; because of one's proximity, they appear only slightly displaced from their positions. As one gradually retreats one will be surprised to find that the red image only appears to be displaced to the extent that the screen is, whereas the violet, blue, and green — the higher images — shift gradually down toward the red. As soon as they have combined with it, it loses its colour, although not completely, and becomes a somewhat roundish, separate image.

---

[22] See Figure 20 in the Appendix.

145.

If one now observes what is happening on the intercepting board one finds that here, too, the eye perceives a contraction of the elongated coloured image, that the violet border no longer seems to play upon the same opening, that the blue, green, and yellow disappear, that finally the red, too, is annulled, the eye perceiving only a white image on the intercepting board. If one moves even further away, this white image assumes colours in the reverse sequence, as has already been adequately described.

146.

Observe, now, what happens on the main screen. The sole remaining image, which is still somewhat reddish, also begins to turn a strong red in its upper part and blue and violet below. Because of this inversion the vanished images of the upper part are unable to re-establish themselves separately. It is on the lower part, the only part remaining, the base, the nucleus of the whole, that the colouration occurs.

147.

Anyone intimately acquainted with these mutually complementary experiments will immediately appreciate the significance of the two horizontally adjacent images (140) and of their displacement, and why it is that the violet must shift from the line of the red without also giving evidence of diverse refrangibility. For, just as all that is true of the whole image must also be true of its several parts so, also, what is true of two adjacent images is also true of their parts; by describing and developing the Newtonian arrangement we shall now give a fuller and more incontrovertible demonstration of this.

148.

A thin strip of white paper, about a finger's width, is fastened across a frame and placed in a dark chamber so that it has a dark background; the red colour from one of the two adjacent prisms is allowed to fall upon it, together with the violet or blue from the other prism; this strip is then viewed through a further prism: the red persists in the same place, being displaced together with the strip, and simply becomes a more fiery red. The violet, however, leaves the paper and moves as a spirit upon the darkness further below, and yet also as a perfectly distinct strip. Once again, this is an appearance to be highly recommended to those still intending to cultivate Newtonian jugglery, well worthy of attention from those now taking their place on the academic bandwagon.

149.

In order that we may switch from astonishment to observation, let us now concern ourselves with the following arrangement: the strip in question should not be too long, simply long enough to accommodate both parts of the image so that each occupies half of it. The sidepieces of the frame to which the strip is fastened should be fairly wide, so that when the images are divided lengthwise their other halves can appear on them. One will then be able to see both images, in all their shadings, one higher the other lower, on both sides of the frame, at one and the same time. The separate parts of both sides, the yellow–red and the blue–red, for example, may now be observed at will on the paper strip. One now experiments as follows: if this arrangement is viewed through a prism, one is aware at one and the same time of the alteration of the entire images and of their parts. While the upper image, which imparts the red colour to the strip, contracts, red maintains its position on the frame and the red colour does not quit the strip. The lower image, however, which imparts the violet colour to the strip, cannot contract without the violet quitting its position on the frame, and therefore also on the paper, on the frame one can still make out the relationship with the other colours, whereas beside the frame the part which has shifted down off the paper appears to be hovering in the air, the darkness behind it acting as its screen, just as the frame does for the changing objective image which is cast upon it. One is able to ascertain with the greatest precision that this is the case, since the isolated downward-shifting coloured strip keeps pace with the same colour in the semi-spectrum at the side, remaining horizontal with it, and moving down and finally disappearing together with it. In order that those who wish to repeat our experimentation and follow us in observing and considering the matter closely may rest fully assured that what has been maintained here is in fact the truth, we shall devote a figure to this arrangement and appearance in Plate 12 of this work.[23]

150.

Having progressed thus far, we now know how the tests which Newton appended to his seventh experiment are to be analysed and assessed. We shall carefully collate them, however, and number them in order to facilitate future references to them.

151.

It is of prime importance here to call to mind the fifth experiment, in which two crossed prisms forced the spectrum to make a bow. This was supposed to prove the diverse refrangibility of the various rays but, according to us, simply gives expression to a general law of Nature, the effect along the diagonal of two equal forces acting at right angles to each other.

152.

We can also carry out the experiment in question so that it combines the subjective and the objective, and to this end we suggest the following experiment. First, by means of a vertically oriented prism, the elongated image of the sun is cast sideways on to a screen so that the colours come to stand next to one another horizontally; the second prism is then held horizontally in front of the eye in the usual manner. The red end of the image will then retain its position whilst the violet extremity will appear to quit its place on the screen and exhibit a tendency to shift down diagonally. Thus prepared, one proceeds to two of the demonstrations proposed by Newton.

153.

VIIa. Another vertical prism is added to the one just described and is positioned so that there are two elongated coloured images in a row.

---

[23] never published.

Once again, these are then observed together through a horizontal prism; both will be seen to have inclined towards the diagonal in such a way that the red end remains stationary, constituting, as it were, the axis around which the image revolves. All of which tells us nothing more than we know already.

154.

VIIb. Nevertheless, this experiment is still worth diversifying. Position the two vertical prisms so that their images are superimposed upon each other in the reverse sequence, the yellow–red of one falling upon the violet of the other and *vice versa*. To the naked eye these images are now coincident but, when they are viewed through the horizontal prism, they appear to shift across each other, each being moved diagonally in its own way. In actual fact this test, too, is simply a curiosity; for, as we shall see, slightly modified conditions produce no variation in it. The same is true of the two that follow.

155.

VIIc. The red and violet parts of both prismatically coloured images are allowed to fall upon each other on the strip of white paper (149); their intermingling brings forth a purple colour. If one then observes this strip through a prism, the violet will move down, detaching itself from the yellow–red, the purple will disappear, but the yellow–red will appear to remain stationary. This is the same as what we have already seen on adjacent images (149), and for us it is proof not of diverse refraction but simply of the determinability of a coloured image.

156.

VIId. Two small round discs of paper are placed a short distance apart; a prism casts the yellow–red part of its spectrum on to one disc and the blue-red part on to the other; the background is dark. Thus illuminated, these discs are then viewed through another prism, held in such a way that the refraction is in the direction of the red circle; as one moves back, the violet shifts closer to the red, finally coinciding with it and eventually moving beyond it. Anyone able to employ the apparatus described will also have no difficulty inflicting and assessing this phenomenon.

All these tests appended to the seventh experiment, like the seventh experiment itself, are simply variants of the main experiment (O. 350–357), which combines the subjective with the objective. For, it is all the same if I make the whole or part of the prismatic image projected on to the wall contract into itself, or if I force it into a diagonal bow. It is all the same whether I do this with one or with several objective prismatic images, whether I carry it out with whole images or with the parts, whether I direct and force them so that they are adjacent, superimposed, crossed, or partly overlapping: the phenomenon remains on one and the same, and it demonstrates no more than that I am able to reduce or annul an objective image brought forth in a certain way — for example, by being directed upwards and then, by means of a subjective refraction applied in the opposite direction, namely downwards, thus reversing its colouring.

157.

It is evident from all this that the parts of a coloured image which have arisen objectively are in fact totally unsusceptible to subjective experimentation, the reason being that, in cases such as this, the entire appearances as well as their parts undergo alteration, nothing of them remaining the same even for a moment. What has been expressed above will now be explained more fully and made perfectly clear by giving an example of the sort of complication that occurs in such experiments.

158.

If this paper strip is illuminated in the dark room with the red part of the image and then observed at fairly close quarters through a second prism, the colour, rather than leaving the paper, becomes much more vivid at its upper edge. What is the origin of this more vivid colouring? Simply that the strip is now acting as a bright red image and, since it acquired a similarly designated upper edge through subjective refraction, appears to be heightened in colour. If the strip is illuminated with the violet part of the image the result is quite different for, although the subjective action certainly causes the violet colour to shift away from it (148, 149), the strip retains the brightness to a certain extent. In the dark room it, therefore, has the appearance of a white strip on a black background and, whilst it assumes colour in accordance with the known law, the violet shimmering which has shifted lower down also plays quite distinctly before the eye. Here once again Nature is wholly consistent, and will yield considerable delight to anyone who has followed our didactic and polemical expositions. The same was to be observed in experiment VIId.

159.

In the other instance of this, which was described above (144), we draw down the separate coloured images subjectively so that they appear as superimposed upon one another. It is only the coloured shimmerings which leave their place; but the brightness they have stimulated on the white screen cannot be removed. These bright colourless images remain behind, therefore, and become coloured in accordance with the known subjective laws, so that anyone who is not familiar with this appearance will find the phenomenon particularly confusing.

160.

We shall now introduce an experiment which relates to the foregoing, and especially to our paragraph 135. We have a shutter with two small round holes which are horizontally adjacent. A blue glass is placed over one and a yellow–red glass over the other, through which the sun shines. Here, as there (135), therefore, we have two differently coloured adjacent images. These are now intercepted by a prism and projected on to a white screen. In this case there is no inequality of upward displacement, they remain below on the one level; close inspection reveals two prismatic images under the influence of the differently coloured glasses, however, and to a certain extent they are modified in accordance with the doctrine of apparent mixing and communication.

161.

The spectrum originating from the yellow glass has shed its upper violet rim almost completely, whereas the lower yellow–red border appears with redoubled vividness; the yellow middle has also intensified to a yellow–red, and the upper blue border has become greenish. The spectrum originating from the blue glass has retained the whole of its violet rim, however; the blue is distinct and vivid, the green moves down, and a kind of purple appears in the place of the yellow–red.

162.

Whether one carries out these two experiments together or one after the other, one is convinced of the inappropriateness of Newton's procedure of operating at random with transient physical colours or with fixed chemical colours. Here, we need only observe that due to their natural differences these colours are bound to bring about totally different results.

163.

It is also instructive to carry out the objective-subjective experiment (O. 350–354) with these two differently coloured prismatic images. Placing the prism before one's eyes, one begins by observing the spectra at close quarters and then gradually moves away from them; both, but particularly the blue, will then withdraw inwards from the top, one finally appearing yellow–red, the other completely blue and, as one moves even further away, they become coloured in the reverse sequence.

164.

The construction described earlier (O. 284) may be mentioned in this connection. Several squares of coloured glass are fitted into a piece of pasteboard and illuminated by means of sunlight, or even only in daylight; by observing them through a prism, one can then indicate precisely what is seen if they are used in order to perform the subjective experiment. It is worth remembering this construction here since, from now on, it will be the main means for investigating the subjective displacement of coloured images: once it has undergone a certain amount of modification and development it will enable one to dispense with most other apparatus.

165.

Finally, the squares which are cut out of the pasteboard are measured with extreme accuracy in order to make perfectly certain that they are all of the same size. The coloured glasses are then placed behind them and observed with the naked eye against the grey of the sky. Since the yellow square is the brightest it will appear to be the largest (O. 16). Whilst the green and blue squares will not yield much to it in this respect, the yellow–red and the violet, which are the darkest, will have the appearance of being quite decidedly smaller. That the relative brightness or darkness of colours should have this physiological effect has only incidentally acquired the distinction of being of prime importance where natural phenomena are concerned.

166.

A prism is now raised to the eye and this series of juxtaposed images is observed. Whereas the images of the spectrum can be altered or even removed, these are specified and fixed chemically and are therefore by their very nature durable; alteration only takes place in the enhancing or curtailing effect of their fringes.

167.

Although anyone can erect this simple construction and although we have already paid repeated attention to these phenomena, we have a particularly significant situation here, which requires that we should give a brief but precise description of the matter. On the yellow image the bright red upper edge is clearly visible, the yellow border merging into the yellow surface; whilst a green develops on the lower edge, although blue, which tends towards a downward-diffusing violet, is also quite distinct. It is much the same with the green image, although it is duller and more subdued, with less yellow and more blue. On the blue image the red edge appears to be brownish and heavily deposited, the yellow border creates a kind of dirty green, the blue edge is greatly enhanced and appears to be almost the size of the image itself. It terminates in a vivid violet border. Since these three images, yellow, green, and blue, appear to form a downward gradation, an inattentive person might regard them as lending support to the doctrine of diverse refrangibility. But, we now have to consider the curious phenomenon of the violet image, which we have already touched upon (45). The yellow–red edge does not clash with the violet; for in apparent colourings yellow–red and blue-red give rise to purple. Now, since the tint of the transparent glass is also of a high degree of purity, it combines with the yellow–red edge to which it gives rise; a kind of brownish purple is formed and, whilst the violet remains undisplaced, together with its upper boundary, there is a vivid and very extensive downward diffusion of the lower violet border. It is, moreover, self-evident that the yellow–red image should be enhanced on its upper boundary and thus remain in line, whilst on its lower boundary all that is visible is a certain smudginess; due to the clash no blue and hence none of the violet which derives from it is able to emerge there.

168.

This test may be diversified by making use of coloured window panes and attaching pasteboard figures to them. If these are placed in front of the sun so that the figures appear dark against the coloured background, one will be able to observe not only the inverted edges and borders but also the ways in which they mix with the colour of the glass. In fact, the more this construction is diversified, the more clearly the friends of truth, candour, and straight thinking will become aware of the falsehood of the first Newtonian test and of all the other experiments relating to it.

Eighth experiment

169.

The author allows the prismatic image to fall on to a printed sheet and, as in the second experiment, proceeds to project this chromatically illuminated script on to a white screen by means of a lens. As in the former case, he then claims that he saw the letters distinctly nearer the lens in the blue and violet light and further from the lens in the red light. We are already familiar with the conclusion he draws from this and since, apart from it being concerned with apparent colours, this is simply a repetition of the second experiment, anyone who casts his mind back to what was involved there will be able to envisage the general outlines of the present procedure. Certain additional factors are involved now, however, and it is advisable, therefore, that this experiment should be submitted to careful scrutiny, too, which should be carried out in the way we found suitable for the second experiment, in order to make perfectly clear the extent to which the two tests run parallel or diverge.

170.

(1) The experimental situation (54–57). In this case the black threads are replaced by printed characters; since they are more or less glazed over by the apparent colours, however, this is not an improvement. What is more, in this as in the former case, Newton overlooks the main point, namely, that the various colours of the spectrum differ in brightness. For the sun's prismatic image divides into two, into its diurnal and nocturnal parts. Yellow and yellow–red appear in the former and blue and blue-red in the latter. The illuminated print shows up most distinctly in the yellow colouring, less so in the yellow–red; for this is already more condensed, and darker. Blue-red is transparent, weakened, providing little illumination. Blue is more condensed, thicker, makes the letters darker or, rather, its dimness transforms the black of the letters into a fine blue, so that they make less of a contrast with the background. It is, therefore, in accordance with these various effects in the various positions that there appears to be a variation in the distinctness of this chromatically illuminated print, this object.

171.

Apart from these deficiencies in the image to which it gives rise, the Newtonian apparatus is unsatisfactory in several other respects. We have, therefore, re-designed it along the following lines: taking a frame which fits our stand (69), we cover it with a sheet of silk paper and, using a concentrated Indian ink, draw various lines, dots and similar calligraphs on the surface. By applying a thin oil to the paper we then render the background to what we have drawn transparent. This screen is used just as the object was in the second experiment. The prismatic image is formed on it from behind, the lens is directed towards the room, and the second screen, on to which the image is to be projected, is positioned at the required distance. Such an apparatus had definite advantages, for it enables this test to be compared with the second one, even to the extent of the shadow lines appearing pure black, free from their glazing effect of the prismatic colours.

172.

Here, once again, we cannot help noticing the total inadequacy of Newton's experimental procedure, the glaring inconsistency of the random way in which he employs his apparatus; making use as he does of whatever comes to hand, the complication and excessive elaboration turn out to be too much for him.

173.

In this connection it has also to be observed that, although what Newton is investigating is alterable, he treats it as if it were not so, as if it were as unalterable as the object of the second experiment. The image which appears here on the reverse side of the transparent paper can, of course, be viewed through an inverted prism, reduced to the point of nullity and then totally reversed, but we shall not dwell upon this here since, for an understanding of the experiment in question, it is convenient to make the temporary assumption that it is a fixed image.

174.

(2) The illumination (57). Apparent colours bring their light with them; they have it in and behind them. Nevertheless, since the different parts of the image are to a greater or lesser extent illuminated

according to the nature of the colours, this image of the print, coloured over as it is, is extremely uneven and defective. Taken as a
whole, this experiment resembles the second in that it is to be classified with the *camera obscura*. As is well known, all the objects to be
projected into the darkened room have to be fully illuminated. But,
both the Newtonian apparatus and our own give rise, not to an illumination of the object, letters or lines, but to a darkening and to an
uneven darkening at that, the shadows of the letters and lines, having
to display themselves in their entirety, do so unevenly in brighter or
darker half-shadows and half-lights. Yet, it is not difficult to convince
oneself that in this respect, too, the redesigned apparatus has considerable advantages.

175.

(3) The lens (58–69). The one we use is the same as that employed
in the second experiment, and this applies to the rest of the apparatus
described there.

176.

(4) The image (70–76). Since the Newtonian procedure begins with
an uneven and indistinct object, how can there ever be a distinct image
of it? As one would expect from the way in which we have characterised this procedure, Newton also runs true to form in making a confession which nullifies the outcome of his test. He begins by assuring us
that he has performed his experiment during the summer when the
light of the sun is at its brightest; he ends with a complaint and an
excuse, in order that we should not be surprised if the test is not particularly successful when repeated. His actual words are as follows:

177.

*For the coloured Light of the Prism, by the interfering of the Circles
described in the second Figure of the fifth Experiment, and also by the
Light of the very bright Clouds next the sun's Body intermixing with
these Colours, and by the Light scattered by the Inequalities in the Polish
of the Prism, was so very much compounded, that the Species which those*

*faint and dark Colours, the indigo and violet, cast upon the Paper were
not distinct enough to be well observed.*

178.

The harm caused by such reservations and qualifications permeates the
whole work. The author starts off by assuring us that he has taken
the utmost care with his preparations — waited for the brightest
conditions, hermetically light-proofed his room, selected the finest
prisms; he then takes refuge behind what is fortuitous, however,
clouds being before the sun, imperfect polishing giving rise to an
unreliable prism. No mention is made of homogeneous and never to
be homogenised lights, which blur and adulterate, invade and disturb
each other — never being and never being capable of being what they
are supposed to be. We have often been reminded of the well-known
theatrical hetman of the Cossacks, for he would have made a first-rate
Newtonian. One can just imagine the self-assurance of the declama-
tions; "If I say a circle, I mean what is not round; when I say homogeneous this always means composite, and when I say white, this
cannot in fact mean anything but dirty".

179.

If we now observe this appearance with our apparatus, we shall discover that the distinctness or otherwise of the black figures is not
related to the colours themselves, but to their degrees of brightness
or darkness, and that in degrees of distinctness the colours are ranged
as follows: yellow, green, blue, yellow–red, and blue–red; the nearer
the last two colours are to the edge, to darkness, the less distinctly
they exhibit the figures.

180.

What is more, one can ascertain a certain point at which the screen,
which intercepts the image parallel to the lens, displays it together
with its component parts with maximum distinctness. One can, however, move the lens towards or away from the object to the extent of
almost a foot without the image becoming noticeably indistinct.

181.

This was the context in which Newton was operating, and nothing is more natural than that when the figures illuminated by the brighter prismatic colours are already to be seen or are still visible, those which are illuminated or, rather, shadowed by the darker colours should disappear. Newton asserts, however, that the letters illuminated by the colours of the diurnal aspect become indistinct[24] whilst those shone upon from the nocturnal aspect are clearly visible. This is quite untrue, as it was in the case of the second experiment, so that nothing Newton attempts to base on this assertion has any validity.

182.

(4) The conclusion. There is nothing further to be added to all that has been advocated and demonstrated in refutation of it.

183.

Before we move away from the context of this experiment, we should perhaps mention some further tests we were encouraged to undertake in connection with it. In order to demonstrate the second experiment as strikingly as possible we replaced the object with differently coloured discs which were strongly illuminated from behind. We then discovered, as might have been expected, that the degree of distinctness exhibited by the dark images which displayed themselves on these discs through cardboard cut-outs or by other means, was also related exclusively to the degree of brightness or darkness in the background. It was this experiment which gave us the idea of replacing the object with painted window panes and, once again, the result was precisely the same.

184.

From here, it was a natural transition to the appearances associated with the magic lantern, which are essentially related to the second and eighth of Newton's experiments; throughout this whole field of research there is unmistakable evidence of the truth of Nature and the

---

[24] Newton means "indistinct" in the sense of "blurred" or "out of focus", not "less contrasted".

naturalness of our exposition as well as the falsehood of the artificial conceptualising of the Newtonians.

185.

We also took the opportunity provided by a portable *camera obscura* to observe a colourfully attired crowd promenading in the brightest sunshine on a public holiday. As soon as people entered or neared the point of distinctness all the variegated clothes in the corresponding region became distinct; all the patterns showed up clearly, the only features that might change being brightness and darkness or both of these with colour, or colour with colour. We therefore venture to reiterate here once again that all natural and artificial vision would be impossible if the Newtonian theory were true.

186.

The principal misconception, incontestable evidence of which is provided by the eighth as well as the first two experiments, is that when coloured surfaces, colours, are appearing and functioning as artists' materials, a property is ascribed to them by virtue of which, after refraction, they sooner or later attain a degree of distinctness; since there can be no point of distinctness without an image, however, and since the aberration which is displayed when an image is displaced by refraction only occurs at the edges, the middle of the image is only affected in exceptional circumstances. Diverse refrangibility is, therefore, a fiction. What is true, however, is that refraction, rather than having a uniform effect upon an image, produces a double image, a sufficient clarification of the property of which has been given in our *Outline*.

Recapitulation of the first eight experiments

187.

We have now reached a point in the unfolding of our polemics at which it is advantageous to call a halt and look back along the path we have taken; let us, therefore, survey the ground we have covered and touch upon the outcome.

188.

Newton's theory, well-known as it is, reiterated by others and by us to the point of wearisomeness, is supposed to have been proved by these eight experiments. And he certainly did what had to be done; for there is little that is new in what follows; he simply uses further means to strengthen his arguments. He diversifies the experiments and constantly performs them under new conditions. He never does anything other than move in a very limited circuit, however, shifting around the same old household effects, first one way then another. As we have got the measure of the eight experiments we now have behind us there will be little that will still be unfamiliar to us in what follows. This is why it has been possible to be so laconic about the dissemination of Newton's doctrine in our compendia of experimental physics. Each of the aforementioned experiments will now be considered separately.

189.

In the third experiment the main phenomenon, the prismatic spectrum, is erroneously treated as a scale, for initially it consists of a duality which only subsequently unifies. In the fourth experiment we meet this appearance in its objective mode — without being provided with any deeper insight into its nature. In the fifth, the image in question becomes somewhat lengthened to one side as a result of a repeated refraction. We have given a comprehensive explanation of the origin of this deviation towards the diagonal, and the elongation.

190.

The sixth experiment is the so-called *experimentum crucis*, and this is perhaps the place to explain precisely what is meant by this expression. *Crux* is taken here to mean a cruciform signpost at a crossroads — this experiment is supposed to prevent us from going awry, to indicate the direct route to the goal. Those who have followed our narrative know what, in fact, it leads to: we arrive at a total impasse which cuts off any kind of progress and does not even open up a prospect. For it is basically simply an *idem per idem*. If the entire prismatic image is refracted in the same direction a second time, it is certainly elongated, but the various colours are no longer distributed in the same way. Thus, what happens to the whole also happens to the parts. Within the whole the violet shifts much further forward than the red, and the separated violet does precisely the same. This is the key to the matter, the false interpretation of which has given rise to so much self-indulgence, so much error. Similar subjective effects are demonstrated in the seventh experiment, and we have traced them back to their true origin.

191.

Although up until then the author had busied himself with extracting coloured lights from sunlight, at an even earlier stage he affirmed that corporeal colours also emit such coloured particles of light: this being investigated in the first experiment, which illustrated the apparent difference in the displacement of coloured squares on a dark background. We have demonstrated in detail the true relationships at play here, showing that the cause of the appearance is simply the action of the prismatic edges and borders on the boundaries of the images.

192.

In the second experiment smaller images were placed on these variously coloured surfaces and, sooner or later, when these images were projected on to a white screen through a lens, their outlines became more distinct. Here, too, we have analysed the true nature of the relationship in detail, as we did in the case of the eighth experiment, which involves prismatic colours and ought to complement the second and put it beyond all doubt. We are confident, therefore, that we have exposed the captiousness and faultiness of the experiments as well as the invalidity of the conclusions drawn from them.

193.

In the course of our doing so, we have constantly referred to our *Outline*, where the phenomena are adduced in a more natural order.

We have, moreover, taken careful note whenever Newton introduces something without preparation with the intention of taking the reader by surprise. We have been no less intent upon both simplifying and diversifying the experiments, in order that one might be able to assess them properly by seeing them in the right light and from a variety of viewpoints. We leave it to the well-intentioned reader to bear witness to whatever else we may have done and achieved in attaining our goal.

Proposition III. Theorem III

The sun's Light consists of Rays differing in reflexibility, and those Rays are more reflexible than others which are more refrangible

194.

Now that the author believes he has fixed firmly in our minds the idea that the pure, white, simple brightness of the light we know is, in fact, a mixture of various coloured lights differing in degrees of darkness, and has convinced himself that he has forced these inner constituents out by means of refraction, he considers the possibility of carrying out this operation by another means, by making use of related situations in order to compel light to reveal its charms.

195.

Refraction is so closely related to reflection that it cannot occur without it; since, therefore, reflection is so potent, why should it not exercise its power upon unviolated light? We have diverse refrangibility; how marvellous it would be if we also had diverse reflectibility. Who knows what the further consequences of this might be! Although the author sets about demonstrating such reflectibility by arranging experiments, it is clear enough that he will have as little success in convincing a forewarned observer as he has had with his other demonstrations, and we for our part will now do all we can to bring out the unacceptable nature of this move.

Ninth experiment

196.

We ask our readers to consult the *Opticks* and see for themselves how Newton now sets to work; for, instead of getting involved with him and pulling him up step-by-step, we now intend to go our own way in presenting the true nature of the phenomenon. To this end, we have taken Figure 28 of Newton's fourth plate[25] as the basis of our eighth plate: we have, however, redrawn it in lines more representative of the nature of the situation and, in order to present the phenomenon more effectively, we have shown it as it gradually develops by employing five figures, so that the observer can see what is taking place. In order that the whole may be more easily grasped we have retained the lettering on Newton's plate: there should, therefore, be no difficulty in comparing the figures. We shall now proceed to explain our copper plates, our intention being to provide further evidence of the general inadequacy and captiousness of Newton's figures as we do so.

197.

Let us take our eighth plate and consider the first figure. The sun's image enters the dark room at F and passes through the right-angled prism ABC to the base M; from there it passes on through, is refracted, coloured, and displays itself in the usual manner on the surface underneath as the elongated image GH. There is, therefore, nothing in this first figure with which we are not already thoroughly familiar.

---

[25] See Figure 21 in the Appendix.

VIII
5
B
F
A
M
C
X
Y
P
t
G
H
N
V
1
B
F
M
A
C
G
H
B
F
M
A
C
Y
X
N
P
t
V
4
2
B
F
M
N
A
C
N
B
F
M
A
X
C
Y
N
P
t
V
3

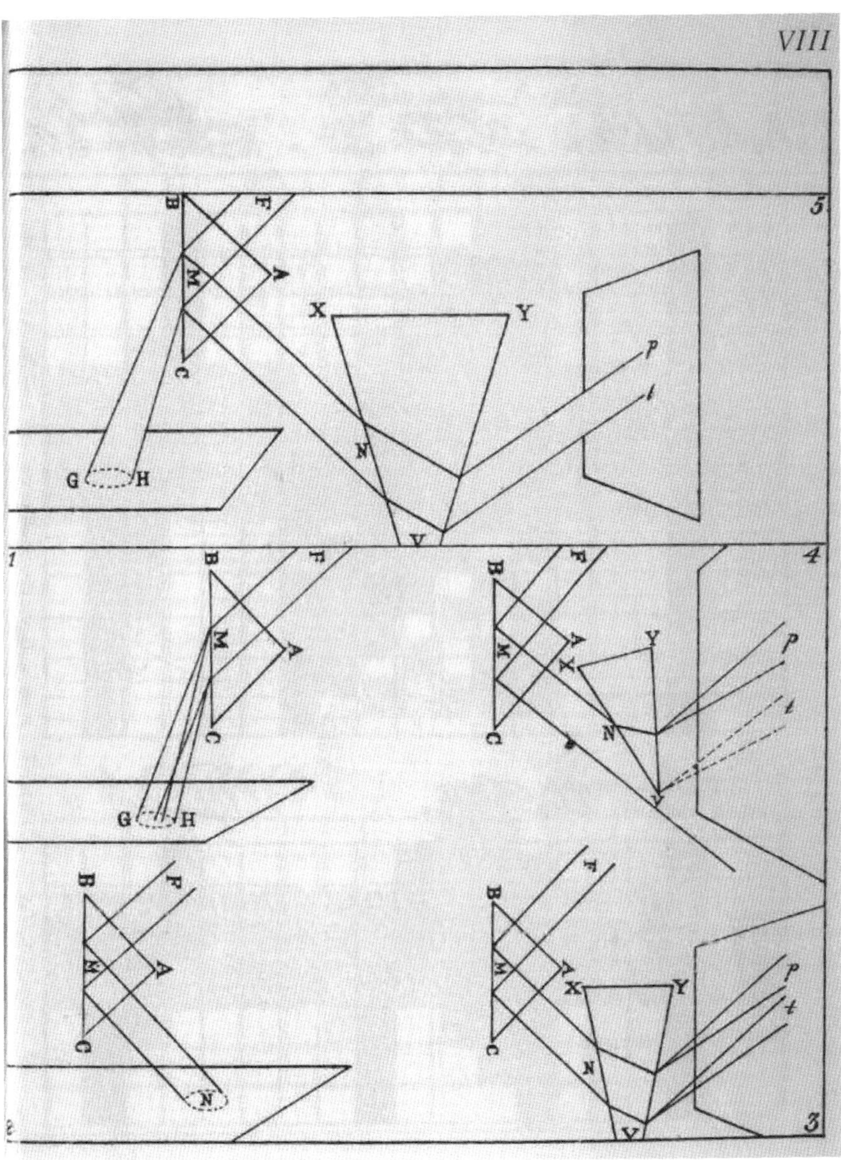

Plate VIII

198.

In the second figure, too, the sun's image enters the dark room at F and passes into the right-angled prism ABC; it is, however, now reflected by the base M in such a way that, after emerging from the side AC, it passes to the surface underneath and projects on to it the round and colourless image N. Although this round image is certainly a derived one, it is unaltered; it has not yet undergone any chromatic determination.

199.

As the third figure shows, this image N is now allowed to fall on to a second prism VXY so that, on passing through it, it forms precisely what an original image, an image reflected by a mirror forms; it there-fore behaves just as we would expect it to, producing the elongated, coloured image pt on the screen opposite.

200.

The fourth figure shows what happens when only the upper part of the image N is intercepted by the second prism VXY. At first, only the upper edge of the image pt will display itself in blue and violet on the screen opposite; only somewhat later, when all of the image N has been intercepted by the prism VXY will the lower edge t of the image appear. It goes without saying that this experiment can be performed either with a direct image of the sun or with one reflected by a plane mirror.

201.

The author errs badly here by persuading himself and his supporters that the multicoloured image GH of the first figure is closely related to the colourless image N of the second, third, and fourth figures; for there is in fact no connection at all. In the first figure the sun's image which arrives at M passes through the face BC and, after refraction, is coloured at GH. The image here is very different from that in the second figure, which is reflected from M to N and remains colourless until the second prism projects it chromatically on to the screen at pt in precisely the same way as it would were it derived from a direct beam.

202.

As is evident from the fourth figure, if part of the image is now intercepted by a prism at an extremely oblique angle (200), the result is the same as that observed by Newton when he turned the first prism slowly upon its axis — one of the apparent examples of acuteness and accuracy on the part of our experimenter.

203.

No one will now have any difficulty in realising that the image which is transmitted at M and forms the image GH on the screen has little in common with that which is reflected at M and projected, colourless, to N. This is made even more clear by the fifth figure. This shows clearly that, whereas after refraction the image M gives rise to the yellow and yellow-red edge G, image N after refraction gives rise to the violet edge p and, conversely, that whereas image M gives rise to the blue and blue-red edge H, image N, when it has undergone refraction, gives rise to the yellow and yellow-red edge t; naturally enough, since in the first prism the sun's image F is refracted downwards, and the derived image M is refracted upwards to N. What we have here, therefore, is nothing new, and if any thoughtful observer fails to grasp what he sees in this particular case, it will be because of the casuistries, confusions, and misrepresentations of Newtonianism. In the twenty first figure of Newton's fourth plate, for example, the image is simply represented by a single line, the author being in the habit of referring indiscriminately to the image of the sun, the light of the ray; and yet in this case particularly, as we have demonstrated in Figure Four of our eighth plate, it is of the utmost importance that the appearance should be regarded as an image, occupying a certain space. It would be quite easy to set-up some apparatus in which all the required equipment was fixed to a framework — this being required if one is to bring the phenomenon about by gentle rotation and so provide the friends of truth with visual evidence of the captiousness and inadequacy of the Newtonian experiment.

Tenth experiment

204.

Here, too, what is required is that several figures and a number of pages should be devoted to the refutation of the experiment — an experiment which is closely connected to the previous one. The time has come, however, when we should be able to leave certain things to the reader, allow him the pleasure of sorting out such confusions for himself. We shall therefore rest content, simply recommending that he should consult Newton's text and the figure accompanying it. He will find a detailed description, an illustration and a scholium, the total outcome of which is merely the encumbering of the ninth experiment with further conditions and particularities; the main point is further obscured without there being any improvement in the cogency of the argument.

205.

We have already called attention to the central issue here (201), and our present task is therefore confined to requiring of the observer or critic that he should hold fast to the fact that there is no connection or relationship between the two prismatic images since one results from reflection and the other from passing through a medium; each has to be considered separately, each has its own origin, each is resolved in its own way, and any of the ways in which they are correlated, such as those Newton is so keen to convince us of, have therefore to be regarded as an empty illusion, as wishful thinking.

Newton's recapitulation of the first 10 experiments

206.

Just as we for our part found it necessary and helpful to round off the treatment of the first eight experiments with a general survey, so Newton, in his own way, provides a survey of the first 10; and since we have every reason to examine him closely in this respect, we find ourselves obliged, once again, to take issue with him. In an incredibly complicated passage he so forces together and superimposes what ought to be separated, that only if one possesses the most intimate knowledge of his previous procedure, only by means of the most careful circumspection, will one be able to avoid this trap, which he laid so early on and which he now finally springs. We would therefore ask our reader to once again patiently bear what he has heard in other connections so many times before; for the only way to counter Newton's error, reiterated as it is to the point of wearisomeness, is to re-state the truth with equal persistence.

207.

Now seeing that in all this variety of Experiments, whether the Trial be made in Light reflected, and that either from natural Bodies, as in the first and second Experiment, or specular, as in the ninth.

208.

Newton brings together here, under the heading of reflected light, experiments which have nothing in common, the reason being that he wants reflection to be accepted as equally important and effective as refraction in the production of colour. The action of the reflecting image in the ninth experiment is no different from that of a direct image, and the reflecting has nothing to do with the production of colour. It is, however, necessary to consider the naturally coloured bodies of the first and second experiment in an entirely different way. Their surfaces have specific properties, their chromatic characteristics are fixed, it is the light which reflects off them which renders their features visible and, as on previous occasions, all Newton really does here is manipulate terminology in order to suggest that a reflection of coloured lights from natural bodies, that is to say, of lights elicited from primary colourless light by certain surface properties, subsequently undergoes a diverse refraction. We, however, have better knowledge of what this phenomenon involves, and we can be imposed upon neither by the misleading manner in which these three experiments are described, nor by the way in which they are now lumped together.

209.

Or in Light refracted, and that either before the unequally refracted Rays are by diverging separated from one another, and losing their whiteness which they have altogether, appear severally of several Colours, as in the fifth Experiment.

210.

The numbering of our paragraphs is of particular value in this connection, for without numbering it would now be difficult to pinpoint our references to the fifth experiment. In fact, it is only of central importance where the fifth experiment is concerned — where we have already done all we could to call attention to the way in which this counterfeit point is smuggled in. Here too, note how artfully and surreptitiously Newton brings in what is false. Take note of his creed! Refraction is not all that is required to separate the colours — to extract them from the original white — to make the various unequal rays appear as the various colours; something more is needed: a divergence. Where, before now, has there been any mention at all of this divergence? In the passage referred to, (112), Newton certainly speaks of a refracted and white light which is also circular, as well as being able to appear coloured and oblong; there is, however, a profound silence concerning the way in which one sort of light develops out of or emanates from the other. Our author is astute enough to make no mention of the word divergence prior to this recapitulation, and then only to do so incidentally, as it were, as if it were something which is self-evident. With regard to his doctrine, however, rather than being self-evident, it constitutes a flat contradiction. Consequently, both here and in the previous passage, (112), it is admitted that the appearance of a light or light image which has undergone refraction can be incompletely coloured. But, if this is the case, why do Newton and his adherents regard refraction and the full appearance of colour as one and the same thing? It is worth taking particular note of the first figure on the seventh of our plates, for it is reproduced in every compendium right up to the present day. There are, moreover, many other such diagrams, even extremely detailed ones such as those in

Martin's *Optics*[26] and it is not always the case that refraction and total divergence of all so-called rays are depicted as taking place in the prism? What, then, is a divergence occurring after a completed refraction? It can only be proof of the underhandedness of one's approach, of one's having had to slip in something which one can neither use nor dispense with.

211.

Newton also proceeds underhandedly in the way in which he maintains that the refracted light image is white and circular (112) for, although in appearance it is certainly white in the middle, its edges are coloured and it is always somewhat elongated. That the colours appear only at the edges, that these edges diverge, that they finely overlap and cover the whole image, that this is the key to the whole matter, that this simple phenomenon makes nonsense of Newton's theory, has been re-stated by us to the point of weariness. In this particular connection, however, we shall not forget to observe that there is some excuse for the obstinacy of the Newtonians. The master was in fact well aware of the circumstances which ran counter to his doctrine. He did not suppress them; he did not fail to mention them — he simply played them down, hid them away. Consequently, when the later Newtonians were confronted with such a circumstance, which contradicted the doctrine, they could give the assurance that, although the master had been well aware of it, he had continued to regard his theory as established and irrefutable, and that such things were therefore matters of no importance. Our conviction is that it is not only refraction which is operative, but also divergence. Take note of this, for in the course of this critique we shall have frequent occasion to refer back to it.

212.

Or after they are separated from one another, and appear colour'd as in the sixth, seventh, and eighth Experiments.

---

[26] Benjamin Martin, *New Elements of Optics*, (London, 1759).

213.

To anyone not convinced by our detailed exposition that the three
experiments in question have failed to accomplish or demonstrate
anything at all, we have nothing further to say.

214.

*Or in Light trajected through parallel Superficies, destroying each oth-
er's Effects, as in the tenth Experiment.*

215.

An image of the sun which has passed through parallel surfaces at
right angles is scarcely altered at all, and if it subsequently passes
through a prism, it gives rise to precisely the same appearance as
would a directly incident one. Like so many of the other experiments,
the tenth is no more than an elaboration of a very simple phenomenon, it merely increases the amount of material to be considered and
is certainly out of place in this recapitulation.

216.

*There are always found Rays, which at equal Incidences on the same
Medium suffer unequal Refractions.*

217.

One never finds rays, one merely uses them to explain appearances;
what is found is not an unequal refraction of an image but one which
is not entirely clear-cut, not sharply defined, and we have provided
an adequate exposition of the origin and cause of this. The heart of
our objection to Newton and his school is that they think they have
seen with their eyes that which they have theorised into the
phenomena.

218.

And that without any splitting or dilating of single Rays.

219.

The concept expressed here is quite erroneous. Newton is in fact maintaining that what happens to white light does not also happen to coloured light; this can only be asserted by the unobservant, by someone who overlooks certain subtle differences. We have demonstrated in sufficient detail that there is no difference at all between what white and coloured images encounter in refraction — that both become prismatically coloured at the edges in the ordinary way.

220.

Or contingence in the inequality of the Refractions, as is proved in the fifth and sixth Experiments.

221.

Newton's insight is wholly justified when he maintains that it is not fortuitously but in accordance with a law that refraction gives rise to the appearance of colour. The *History* will make clear to us how this *aperçu*[27] has served as a basis for his errors; it will therefore also enable us to explain quite a number of other matters.

222.

And seeing the Rays which differ in Refrangibility may be parted and sorted from one another, and that either by refraction as in the third Experiment, or by Reflection as in the tenth.

223.

In the third experiment we see the series of colours in the spectrum; to go on to say that this series consists of separated and sorted rays

---

[27] Goethe used this word when he wished to remind the reader that a "rapid glance" at a basic phenomenon of Nature, a *Urphänomen*, is the correct way of attaining a true perceptual encounter with the laws of physics. In other words, first impressions are the most important ones; such as the ones he formed when he first looked through a prism to perceive the boundary colours.

is to do no more than make use of a hypothetical explanatory formula, and we are well aware of how extremely inadequate a formula it is. The only thing that happens in the tenth experiment is that, whilst one spectrum disappears on one side, a new one appears on the other, neither as a whole, nor in its parts, does the second derive anything from the first, there being not the slightest connection between them.

224.

And then the several sorts part at equal Incidence suffer unequal Refractions, and those sorts are more refracted than others after Separation, which were more refracted before it, as in the sixth and following Experiments.

225.

We have dealt so thoroughly with the so-called *experimentum crucis* and with whatever Newton associated with it, and in doing so we have so carefully explained and elucidated the misleading circumstances and hidden factors it involves that all we now have to do is repeat that this experiment, which is supposed to guide us along the right path, provides evidence not of any diverse refrangibility but of a repeated and continuous refraction constantly taking effect in accordance with perfectly simple laws.

226.

And if the sun's Light be trajected through three or more cross Prisms successively, those Rays which in the first Prism are refracted more than others, are in all the following Prisms refracted more than others in the same Rate and Proportion.

227.

Once again we have a cross, and the ordinary human mind is crucified upon it; for the situation is the same here as it is with the *experimentum crucis* in which a repeated and continuous refraction acts in the same plane as the initial one. In this fifth experiment, however, the repeated and continuous refraction acts sideways, the result being that the image is forced into the diagonal and subsequently into a constantly increasing inclination in such a way that, because of the constantly increasing displacement, it becomes more elongated.

228.

It's manifest that the sun's Light is a heterogeneous Mixture of Rays, some of which are constantly more refrangible than others, as was proposed.

229.

All that is manifest to us is that the sun's image, just like any other bright or dark, coloured or colourless image, in so far as it distinguishes itself from its background because of the refraction, acquires a coloured accessory image at its edge, and that this accessory image, under certain circumstances, can extend and cover the main image.

230.

Newton's totally false assumptions naturally prevented him from drawing any valid conclusions. Any attentive reader will certainly agree that his ten experiments have not enabled him to prove anything. The advantage of our having carried out this investigation is, firstly, that we have rid ourselves of an erroneous and empty opinion, secondly, that we have gained a clear conception of the significance of a previously derived phenomenon (O. 178–356), and thirdly that we have become acquainted with a prime example of a sophistical misrepresentation of Nature, which could only have been brought about by an intellect as remarkable as Newton's, the genius of which is matched only by its wilfulness and pertinacity. Having progressed as far as this we shall now attempt to make our polemics more congenial, both to ourselves and to our readers.

Review of what follows

231.

Had we allowed ourselves to be persuaded by Newton's recapitulation, had we been disposed to applaud his words and accept his theory, we would have wondered why he does not regard the matter as settled, why he not only continues to work out proofs but even starts again from the beginning. It is therefore essential, if we are to appreciate the objective he now proposes to reach, that his new point of departure be brought under review.

232.

We shall begin by making a few general remarks. For a long time Newton's doctrine was only known to the world of natural science from his letter to the London Royal Society. The extent to which one was able to investigate the theory was therefore more or less a matter of ingenuity and luck. Mariotte[28] who, since he had investigated the *experimentum cruces,* had probably noticed the extraneous chromatic fringes of the small coloured images in the second refraction, questioned the fundamental principle that the homogeneous lights separated from heterogeneous white light are unchangeable, and yield no other colour than their own when repeatedly refracted. Newton found a way around this by maintaining that the lights separated out by means of the simple prismatic experiment had not been separated enough, and that if they were to be, a further operation was required. The four experiments that follow were devised especially for this purpose; they were designed to deal with this adversary, and were subsequently also employed to this end by Desaguliers.[29]

---

[28] Edmé Mariotte (1620–1684), a prior at Dijon and critic of *Newton*. His *Traité de la Nature des Couleurs* (1686) was translated and used by Goethe. Mariotte's lack of success on repeating the *experimentum cruci*s was due to the fact that the colours were not sufficiently separated before reaching the second aperture; and it was this difficulty which led Newton to introduce a lens in the eleventh experiment.

[29] Jean Desaguliers (1683–1749). Theologian and physicist at Oxford and a Huygenot refugee. He was a Newton protagonist and rejected Mariotte's polemics against Newton.

233.

He begins, therefore, by making further elaborate arrangements for finally and decisively bringing about the complete isolation of the various homogeneous lights which are hidden away in heterogeneous light, and which have hitherto only been imperfectly separated. This is the purpose of the eleventh experiment. He then makes a further attempt to provide precise visible evidence that these lights which have now been well and truly separated remain unaltered when refracted afresh. The twelfth, thirteenth, and fourteenth experiments are supposed to further this aim.

234.

How often have we heard both propositions repeated; with what decisiveness has the author declared the two issues to be settled: and here we are, beginning all over again as if nothing had happened! The self-assurance of the school is increased, since the master has succeeded in presenting and confirming the matter in such a variety of ways, examined more closely, however, his approach is simply the raindrop method, which by constantly dripping on to the same spot eventually wears the stone away, nevertheless, the final effect is no different from that produced by the immediate use of genuine brute force.

235.

In order to highlight the practical implications of this he proceeds once more to emphasise the naturalness of the conclusion to be drawn from this erroneous concept, namely, that in spite of the common point of incidence of the composite heterogeneous light, each homogeneous ray separated out by the refraction has its own distinctive path, so that what was previously united is subsequently irretrievably separated.

236.

In the interest of practicality, and in a manner he takes to be irrefutable, he now deduces from this that there is no improvement to be made in dioptric telescopes. Dioptric telescopes have been improved, however — although very few have been ready to draw the retrospective conclusion that the theory must therefore be wrong. The *History* is particularly interesting in this respect, since it will show us that long after improvements have been made in dioptric telescopes, the school has continued to assert its totally theoretical conviction that they are an impossibility.

237.

So much for the remaining contents of the first part. The author does nothing more than change a few words while repeating his earlier pronouncements, and alter a few of the conditions whilst reproducing his earlier experiments; once again, therefore, we have to arm ourselves with caution and be patient.

238.

Finally, Newton gives an account of the reflecting telescope he constructed, and we have to congratulate both him and ourselves that he should have laboured under such a misconception and still come up with such a genuinely useful solution. One simply has to admit that error, in so far as it gives rise to compulsion, can drive us towards what is true, whilst what is true, when it grips us too violently, can make us far too ready to take refuge in error.

Proposition IV. Problem I

*To separate from one another
the heterogeneous Rays of compound Light*

239.

Why should Newton now attach any further importance to this task? He has already given assurance that the homogeneous rays have been separated from one another (212), parted and sorted out (222)! He knows only too well from the criticisms of his antagonists, however, that he has not really achieved anything, and he admits that only a partial separation has taken place. This accounts for his now launching into an extensive discourse, into the exposition of the:

Eleventh experiment

Into the illustration of the figure pertaining to it, and achieving just as little as he did earlier on; as usual, he simply complicates the matter, so that only the well-informed can see what he is up to.

240.

Now, since all this takes place after the rounding off of the recapitulation, one has to assume that what is now discussed is simply what has already been dealt with. Consequently, if we now followed the main line we have adopted so far and challenged the author word for word, we too would do no more than repeat ourselves and lead our reader back into the labyrinth we have now managed to leave behind us. We have therefore decided to adopt another procedure; we intend to demonstrate that the task undertaken is impossible, and in order to do this we have only to call attention to what has already been demonstrated in detail in numerous other connections — especially that of the fifth experiment.

241.

Everything depends on one's realising that in objective prismatic experiments the sun is to be regarded as no more than a luminous image; on one's also being aware of what happens when a bright image is displaced. On one side there is the appearance of the yellow-red edge which emerges into the yellow as it diffuses inwards towards the bright centre, on the other the appearance of the blue edge which merges into the violet as it diffuses outwards towards the darkness.

242.

The two coloured sides are parted, separated, and divorced from the outset; on the other hand, the yellow cannot be parted from the yellow-red nor the blue from the blue-red. By displacing the image further one can so broaden these edges and borders that the yellow and blue merge, and green arises; the series of colours produced in this way can become so lacking in separation, because of the repeated elongation of the image, that the increasing extent to which the inner colours yellow and blue superimpose eventually gives rise to their forming green, the result being that only three colours remain.

243.

Those who have fully understood the appearance that we have repeatedly examined will have no difficulty in evaluating Newton's approach to it. Newton prepares an extremely small luminous image, and by means of a curious arrangement displaces it to such an extent that he claims that it is 75 times longer than it is broad, we admit the possibility of such an appearance; but what is the point of it?

244.

The elongation of a bright image, be it a large or a small one, is only brought about by the outer violet border; the inner yellow border combines with the blue edge and does not move outside the image. It follows that given the same displacement the relationship between breadth and length in a small image differs from that in a large one; but Newton does not want to admit this, since it directly contradicts his doctrine (90–93).

245.

If one has a good grasp of the matter, the erroneousness of the Newtonian presentation of it, which we have already submitted to the required examination (103–110), is immediately apparent. We shall now add the following: according to Newton the elongated image really consists of intermeshing circles which are, as it were, coincident, superimposed within the white image of the sun, and which are now forced asunder by refraction as a result of their diverse refrangibility. It then occurs to him that if the diameter of these circles is reduced and the prismatic image elongated as much as possible, their separateness and distinctness will be so increased that they will no longer intermesh as they do in the larger image. To visualise this, stack a number of silver *thalers* and an equal number of *groschen* side-by-side on the table, tip them over and carefully spread them out into two parallel lines so that the centres of the *thalers* and *groschen* remain adjacent; one will then have no difficulty in observing that the *groschen* are already separated when the peripheries of the *thalers* still overlap. It was in this crude way that Newton conceived of the diverse refrangibility of his homogeneous rays and this was also the way in which he illustrated it, — see Figures 15 and 23 of the *Opticks* and Figures 5, 6, and 7 in our seventh plate. However, since all this drawing out of the image in both the previous and the present experiment has not enabled him to separate out the colours, he continues to depict these circles by using dotted lines, so that they appear on paper as being separated, and yet, not separated. It also gives rise to his taking refuge behind another supposition and assuring us that there are not five, nor seven homogeneous rays, but an infinite number of them. Consequently, if one undertakes the laborious operation of separating out rays one regards in the first instance as being adjacent, one is constantly being frustrated by the emergence of an intermediate ray, proving the impossibility of what one is attempting.

246.

Although no attempt has been made to check this experiment, it has become common practice to use it in schools in order to illustrate the possibility of completely separating out these supposedly homogeneous rays; what is more, the figures illustrating the hypothesis have been brazenly reproduced without any attention being paid to Nature or to the experiment. In order to give an example of the presumptuousness required of any compiler of a compendium who has to finish off a chapter based on non-existent or erroneous research, we cannot refrain from reproducing word for word what Erxleben has to say in paragraph 370 of his *Physics*.[30]

Coloured light consists of as many circles as the colours it contains. One of which is red, another orange-red etc., and the last of which is violet. These colours flow together into one another in the coloured bands. Each of these circles is the image of the sun, and since this image derives from such a light, the refrangibility of which is diverse, it is also unable to fall in one place. However, it is only because of their size that these circles intermingle, and if they are reduced by holding an upright lens between the prism and the hole in the window shutter, each simple light appears separately in the form of a small round disc, in an overlapping series; in Figure 75, (a) and (b) are the red and violet lights respectively.

In the figure referred to, the seven lights are depicted as seven small circles ranged neatly and tidily next to one another, just as if someone had actually seen them; the connecting dots, which Newton was always clever enough to retain, are omitted. And there the figure has stood, in between other mathematical line drawings and diagrams

---

[30] J.C.P. Erxleben (1744–1777), Professor of Physics at Göttingen. The book referred to was *Anfangsgründe der Naturlehre* (1772).

of various reliable observations, quite safe and sound, through all the Lichtenberg editions.[31]

247.

In order to complete this account of all our reasons for having dealt more summarily with the eleventh experiment than we have with the others, we also have to mention the following: this is the first time Newton combines prism and lens, and he does so without giving us any indication of what actually happens when these two instruments, which are so closely related and yet so very different, are used together. Here he wants to make use of them in order to separate out his imaginary lights; later on he will employ them in precisely the same way in order to re-unite what he imagines he has separated and so reconstitute his white light. It is this latter experiment in particular which has pride of place amongst those always cited triumphantly by his followers. As soon as a suitable opportunity arises, we shall therefore clarify what actually happens when prisms and lenses are combined in the same experiment. When we have done so we shall be able to repeat the eleventh experiment and bring to light its true relationship, which we shall also have occasion to refer to when dealing with Desaguliers' criticism of Mariotte.

---

[31] G.C. Lichtenberg (1742–1799) was a Professor of Philosophy and Physics at Göttingen — and a Newtonian. He edited the 6th to 9th editions of Erxleben's *Naturlehre*.

Proposition V. Theorem IV

Homogeneal Light is refracted regularly without any Dilatation splitting or shattering of the Rays, and the confused Vision of Objects seen through refracting Bodies by heterogeneal Light arises from the different Refrangibility of several sorts of Rays

248.

The first part of this Proposition has been already sufficiently proved by the fifth experiment.

249.

We have given detailed proof that the fifth experiment proves nothing.

250.

And will farther appear by the Experiments which follow.

251.

As will be made clearer by our commentary that this is a groundless and unprovable assertion.

Twelfth experiment

252.

A black paper.

253.

Why a black paper? Any perforated wooden board, pasteboard or tin sheet would be perfectly in order; perhaps black paper is used once more in order to make it appear as if great care has been taken that there should be no disturbance from any other light.

254.

In the middle of a black Paper I made a round Hole about a fifth or sixth Part of an Inch in diameter.

255.

Why was the hole so small? Simply in order that observation might be more difficult and any variation less obvious.

256.

Upon this Paper I caused the Spectrum of homogeneal Light described in the former Proposition, so to fall, that some part of the Light might pass through the Hole of the paper. This transmitted part of the Light I refracted with a Prism placed behind the Paper, and letting this refracted Light fall perpendicularly upon a white Paper two or three Feet distant from the Prism, I found that the Spectrum formed on the Paper by this Light was not oblong, as when 'tis made in the third Experiment by refracting the sun's compound Light, but was so far as I could judge by my Eye perfectly circular, the Length being no greater than the Breadth. Which shows, that this Light is refracted regularly without any Dilatation of the Rays.

257.

Here, once again, the author indulges in one of his ruses. Apart from minor changes in its conditions, this experiment is identical to the sixth; since it is described here as if it were a new one, however, it adds unnecessarily to the number of experiments and, if one is not careful, it is easy to mistake repetition for confirmation, for fresh proof. All that really happens is that error which has already been given expression is forced home, whereas one imagines that one has acquired new grounds for one's conviction.

Thus, our detailed criticism of the sixth experiment applies here, too, and we shall refrain from repeating what has already been reiterated so often.

258.

We do wish to make one further observation, however. Although the author states that he allowed a homogeneous light to pass through the hole and be refracted a second time, he does not mention its colour. No doubt it was red or yellow-red, which would serve him well to conspire with him, as it were, to refuse to reveal what he is so keen on concealing. How different the outcome of the experiment would have been had he performed it with any of the other colours and had the will to note the results properly!

259.

The following two experiments are prismatically subjective. Although our readers will find the appropriate explanation of their significance in the *Outline*, there is no reason why they should not be dealt with again at this juncture.

Thirteenth experiment

260.

In the homogeneal Light.

261.

Once again, in all probability, in the red.

262.

I placed a Paper Circle of a quarter of an Inch in diameter.

263.

Why this tiny disc again? What is to be observed on it? But since we have now got used to working with tiny holes in shutters, why not also with tiny paper cut-outs!

264.

And in the sun's unrefracted heterogeneal white Light I placed.

265.

Take particular note: the matter concerning homogeneous and heterogeneous light is now fully settled. That which was still to be proved is now declared to be decided, determined, defined, and impresses itself ever more deeply upon the mind of the unsuspecting student.

266.

Another Paper Circle of the same Bigness.

267.

No doubt just as small in order that the whole of the surface viewed through the prism might be coloured at once.

268.

Then I retreated a few steps and viewed both Circles through a Prism. The Circle illuminated by the sun's heterogeneal Light appeared very oblong, as in the fourth Experiment, the Length being many times greater than the Breadth; but the other Circle, illuminated with homogeneal Light, appeared circular and distinctly defined, as when 'tis view'd with the naked Eye.

269.

As has already been indicated above, this latter circle was probably also illuminated by red light and, since Newton himself uses coloured paper instead of prismatic colours in the first experiment, we can simply refer our readers to the comments that have been made in connection partly with the sixth and partly with the first experiment. Consider our third plate where, amongst other squares, there is a red one and a white one on a black background; view these squares through a prism and then read the description we have already given (O. 271–272); one will then grasp the origin of the illusion by which Newton deceived himself, and indeed wanted to do so permanently. When he goes on to say that this experiment:

270.

Proves the whole Proposition.

271.

No one who is better informed will agree; they will, rather, recognise the familiar error, and if they have ever entertained it, rid themselves of it permanently.

Fourteenth experiment

272.

In order that our readers may make an immediate assessment of the value of this experiment, we have designed a plate of six squares illuminated by the primary colours, and drawn on these squares various dark, light, and coloured bodies. View this plate through a prism and then read Newton's interpretation of the resultant appearance; it will certainly be apparent that it is and can only be dark bodies that he is observing in the so-called homogeneous light. Our experiment, however, offers a variety of examples, and it is only by this means that a complete and clear insight into the phenomenon is to be achieved.

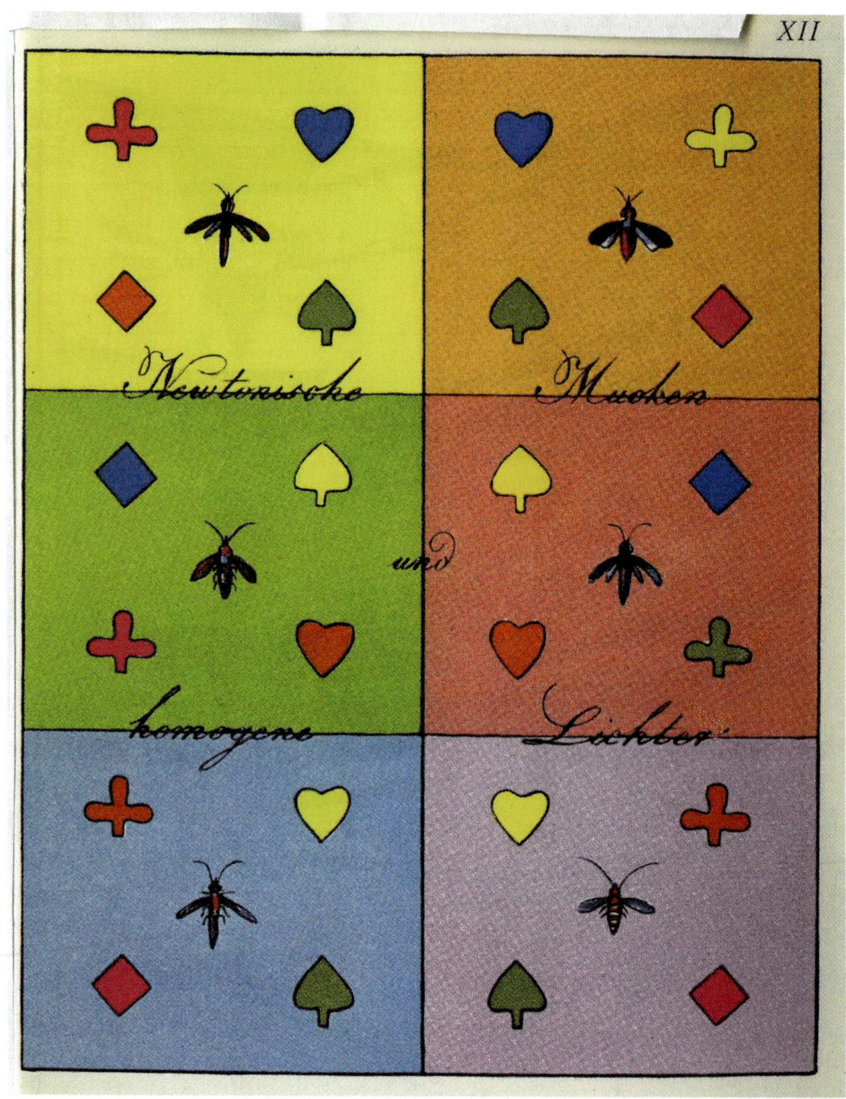

Plate XII

273.

In the homogeneal Light I placed Flies, and such-like minute Objects, and viewing them through a Prism, I saw their Parts as distinctly defined, as if I had viewed them with the naked Eye.

274.

The relationship which manifests itself here must already be sufficiently familiar to our readers, especially those who have not remained indifferent to our didactic dissertation. It is that the edges of a coloured image on a dark background, especially when the colours themselves are dark, can only be observed with difficulty. Here we have the converse case. Newton places dark images against a coloured background, and these images are therefore also illuminated, and to some extent tinged by the coloured light produced by the background. It is only natural, therefore, that the prismatic fringes one sees on these objects should appear smaller, should blend with the objects or should be dissolved at the opposite end, and that they should therefore be somewhat limited and lacking in perceptible borders. In our twelfth plate, therefore, in order to demonstrate the phenomenon clearly in all its aspects once and for all, we have placed light, dark, and coloured images on coloured backgrounds. Since they can all be observed through the prism at the same time, one is able to get plenty of practice in recognising and observing how the edges and borders are affected by the various relationships of brightness and darkness and the properties of the different colours. The observer will become aware of the unfortunate nature of the Newtonian approach, which always gives rise to only that phenomenon being selected which can be turned to advantage, to keeping silent about and concealing all the others; and from the beginning to the end of the revered *Opticks* this is the method employed.

It is hardly necessary to translate and explain the rest of the text relating to this experiment, but we shall not complain at carrying out so light a task.

275.

The same Objects placed in the sun's unrefracted heterogeneal Light, which was white.

276.

Note this: black on white!

277.

I viewed also through a Prism, and saw them most confusedly defined, so that I could not distinguish their smaller Parts from one another.

278.

Of course not! For the smaller, narrower parts were completely outshone by the fringes and so rendered unrecognisable.

279.

I placed also the Letters of a small print, one while in the homogeneal Light, and then in the heterogeneal, and viewing them through a Prism, they appeared in the latter Case so confused and indistinct, that I could not read them; but in the former they appeared so distinct, that I could read readily, and thought I saw as distinct, as when I view'd them with my naked Eye. In both cases I viewed the same Object, through the same Prism at the same distance from me, and in the same Situation.

280.

Although it looks here as though the author has paid full attention to the condition he has in fact disregarded the key one.

281.

There was no difference, but in the Light by which the Objects were illuminated, and which in one Case was simple, and in the other compound.

282.

So here, therefore, the variation in behaviour is supposed to stem from the difference between simple and compound but this difference has always been there — it was already embedded there, unproven in the first proposition, and cropped up regularly in every subsequent proposition and experiment.

283.

And therefore the distinct Vision in the former case, and confused in the latter could arise from nothing else than from that difference of the Lights.

284.

Of the lights indeed; not in so far as they are coloured or colourless or compound, however, but in so far as they appear brighter or darker.

285.

Which proves the whole Proposition.

286.

Our most emphatic reply, for which we hope we have the approval of our readers is that it proves nothing at all!

287.

And in these three Experiments it is farther very remarkable that the colour of homogeneal Light was never changed by the Refraction.

288.

What, indeed, is very remarkable is that it is only here that Newton mentions what constitutes the very ABC of observations, namely that a coloured surface is no more altered than a black, white, or grey one. All that happens is that the boundaries of the images exhibit various colours. If, when investigating this, the coloured spectrum is viewed through a prism at fairly close quarters, so that it is not noticeably displaced, and its versatility (O. 350–356) remains unrevealed, it may be assumed for the purpose in hand that the surface it illuminates is really coloured. And, since the author thinks that he has now brought his proof to a happy conclusion (whilst we are convinced that, in spite of all his pains, his work is highly unsuccessful) we now intend to follow-up his further conclusions very closely indeed.

Proposition VI. Theorem V

The Sine of Incidence of every Ray considered apart, is to its Sine of Refraction in a given Ratio

289.

Rather than take issue with the author, we shall present the matter as it is, in accordance with its nature, and then work back to the appearance's first principles. The laws of refraction were first discovered by Snellius. It was then known that for any medium the sine of the angle of incidence is always in the same ratio to the sine of the angle of refraction.

290.

It was customary to illustrate this discovery in a linear drawing — such as we have reproduced in the first figure of our eleventh plate. A circle was drawn and divided by a horizontal line; the upper half then represents the thinner and the lower the denser medium. Both halves were then sub-divided by a perpendicular line. The angle of incidence was shown as meeting the angle of refraction in the centre, thus giving expression to the measure of their inter-relationship.

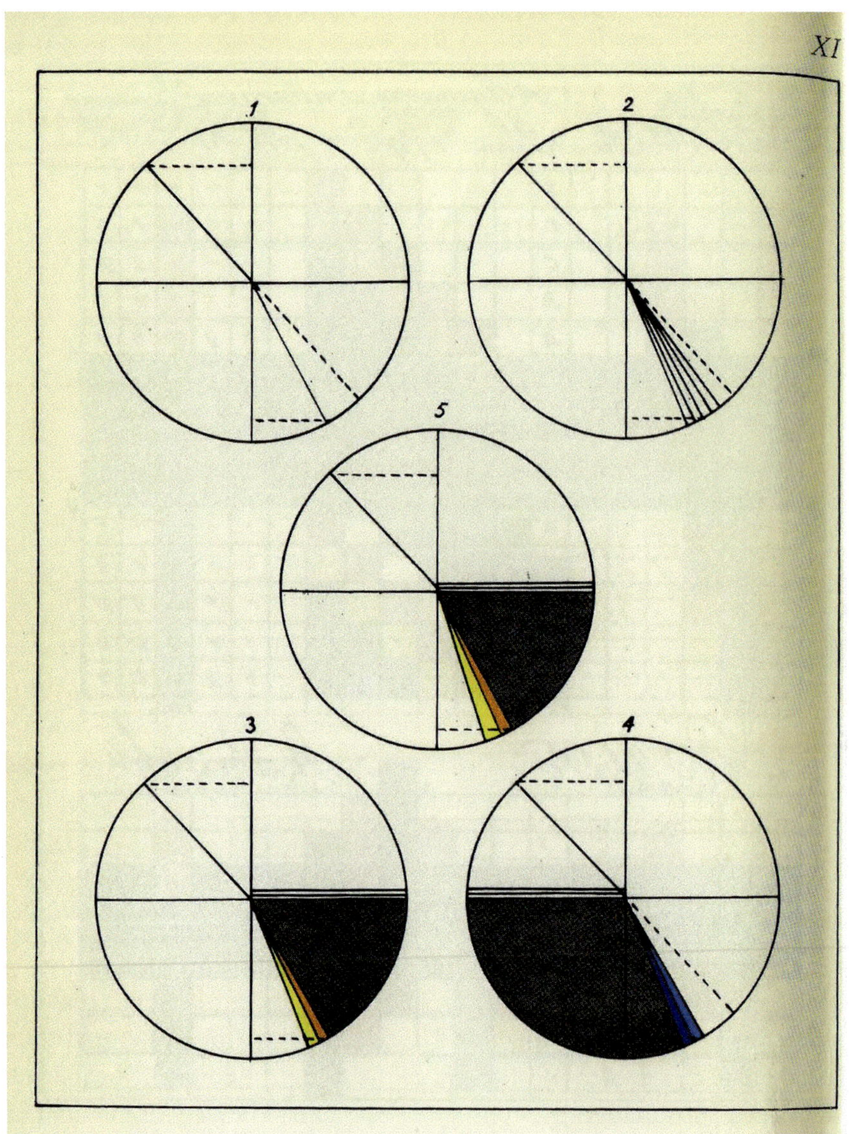

Plate XI

291.

This is a suitable and effective way to illustrate the theory and provide
an *in abstracto* representation of the relationship; but in order to

actually measure both angles against one another some apparatus is needed which is not shown in the figure.

292.

Allow the sun to shine into an empty vessel (O. 187) in such a way that the shadow cast reaches precisely as far as the other side, completely covering the bottom. Pour water into the vessel so that the shadow will retreat towards the side from which the light is coming. Initially, it is the direction of the incident light which is observed, then that of the refracted light. On what evidence do we measure these two directions if it is not by using the shadow, and indeed the shadow's boundary? Consequently, in order to ascertain the measure of the refraction a bounded medium is required.

293.

Let us continue. When the law of refraction mentioned above was discovered, no attention was paid at all to the appearance of colour which accompanied it, and one has to admit that in the case of parallel media this is a very slight effect. The refraction at the point of incidence of strong, white, bright light was measured, observed, and illustrated as described above. Newton then discovered that the refraction is accompanied, in accordance with a certain law, by an appearance of colour, he explained this by supposing that the white light contains various coloured lights which, because of their diverse refrangibility, separate out to appear side-by-side.

294.

It followed naturally from this that if the white light had a particular angle of incidence, let us say 45° here, the angle of refraction of the rays separated out by the refraction must vary, since some were diverted more than others and, therefore, that whereas within incident light there was only one sine, these refractive angles required that five, seven, or rather an infinitude of sines should be taken into consideration.

295.

To make this comprehensible, Newton employed a simplified and less detailed version of a figure used previously to illustrate the relationship between refraction and incidence.

296.

A light ray had been postulated as a matter of convenience, for the abstract replaced millions of rays; no mention had been made of the boundary in the figure in question, for it was taken for granted. Now Newton does not mention a boundary either; he does not take it for granted, however, passing it over, leaving it out, and drawing his figure as it is reproduced in our No. 2.

297.

Consider, however, what has been demonstrated above — particularly by the diagrams. Let the innumerable rays of the sun pass through the upper semi-circle of the thinner medium to the lower semi-circle of the denser medium at an angle of 45°; by what means is one supposed to be able to observe the relationship between what happens after refraction and the light rays now falling on the open horizontal line or surface of the denser medium? How does one propose to discover the relationship between the angle of incidence and the angle of refraction? The first requirement is that one should be able to indicate a point at which the two can meet.

298.

It is quite impossible to do this by sliding any sort of obstacle or screen across one side to the middle, that is to say, either from the side where the light originates, as is shown in No. 4, or from the opposite side, as is shown in No. 3. The ratio of the sine of the angle of incidence to the sine of the angle of refraction is precisely the same in both cases, the only difference being that in the first case the light shifts towards the darkness and in the second the darkness towards the light. In the first position, therefore, a blue and blue-red border appears, in the second a yellow and yellow-red one; there is, moreover, no differentiation in their refraction, let alone any occurrence of a refrangibility.

299.

This is, therefore, a good time to point out that, whereas any universal law of Nature can be expressed by linear symbolism, and that, when so doing, one can certainly presuppose the conditions which cause the phenomenon which relates to the law, one is not justified in using such figures for carrying out further operations on Nature, and when delineating a phenomenon which only occurs under the most specific conditions, it is precisely these conditions which must not be ignored, kept silent about, set aside; one has to take good care that they, too, are expressed in universal terms and represented within the symbolism. We believe that we have achieved this in our eleventh plate, thus putting the coping stone on what has been painstakingly built up in the *Outline* and bringing the matter to its ultimate conclusion. It is to be hoped, therefore, that these figures in particular will be included in future compendia, for they provide the best and most succinct means of introducing and debating the doctrine.

300.

Finally, in the fifth figure, by illustrating the phenomenon which gives rise to achromatism and hyperchromatism, we have managed to present everything on one page. It is assumed that a medium having the same refractive index as the previous one has the power to cause a further extension of the chromatic appearance. Although the incidence and the refraction are the same as those of the first medium, there is a considerable difference in the chromatic appearance. The phenomenon is exhibited here *in abstracto*. But it may also be possible to illustrate it from Nature, for we are already able to augment the chromatic appearance produced by a medium without perceptibly altering its power of refraction. This is, therefore, the place to repeat the conjecture (O. 686) that it may be possible to do away completely with the chemical property whereby a refracting medium gives rise to edges and borders.

301.

Anyone who has had no difficulty in following and grasping what we have presented here, will remain unimpressed by all the measuring, reckoning, and discoursing Newton expends upon this proposition — and all the more so since subsequent observations have long since played such havoc with his antiquated construction. We shall, therefore, pick no quarrel with the:

Fifteenth experiment

302.

We encountered the sideways motion of the spectrum in the fifth experiment, and it is now reproduced by means of a number of prisms, all that is achieved, however, is that the increasingly elongated spectrum becomes increasingly oblique; since all this is sufficiently familiar to us, it adds nothing of interest.

Proposition VII. Theorem VI

The Perfection of Telescopes is impeded by the different Refrangibility of the Rays of Light

303.

Since there are various ways in which one can become involved with a science or approach a particular phenomenon, one's whole treatment of a subject often depends upon one's initial impression of it. If this is borne in mind when considering the history of knowledge, if a precise note is taken of the way in which certain individuals, societies, nations, and contemporaries make and exploit a discovery, much that we should otherwise be unaware of or find puzzling will speak for itself. We shall often have occasion to explore this theme in the *History of Chromatics*, and we shall also find it a help in reaching an assessment of this section of the text now under discussion. First and foremost we take note of the fact that it was as a consequence of his attempting to improve the dioptric telescope that Newton acquired his interest in the theory of colours.

304.

When the laws of refraction were discovered there was good reason to ignore the appearance of colour, for it is of no significance in experiments involving parallel media. When polished glasses were used for spectacles and telescopes, however, the phenomenon became more a matter for discussion. Once telescopes had been invented, mathematicians and technicians got down to the business of improving them, and paid particular attention to two defects which were generally known as aberrations or deviations. One stemmed from the form; it was noted that lenses consisting of spherical sections, rather than concentrating the parts of the image in one distinct point, caused the convergence of the rays (the concept was employed) to take place partly too early and partly too late. It was, therefore, suggested that use should be made of elliptical and parabolic glasses, but the experiments carried out were not entirely successful.

305.

During the course of these investigations the second or chromatic deviation was noticed. It became evident that an appearance of colour reduced the distinctness of the images, particularly through the blurring of their boundaries, which are of primary importance to their clarity. For a long time this appearance was regarded as incidental, and was simply ascribed to various fortuitous circumstances such as irregular refraction or flaws in the glass; effort was therefore concentrated upon the neutralisation and elimination of the first kind of deviation, that stemming from the form.

306.

Newton, on the other hand, concentrated his attention on the second type of aberration. He finds that the chromatic appearance is constant, and since he is working on the basis of prismatic experiments, that it is extremely well-defined; he firmly convinces himself of the doctrine of diverse refrangibility. We have seen how he backs this doctrine up; the *History* will make it clear to us how he was misled into doing so.

307.

According to his observations, the way in which he interprets them, the manner in which he theorises, the inference drawn from the proposition is perfectly correct; for if colourless light is diversely refrangible the appearance of colour cannot be separated from refraction, the aberration cannot be neutralised, there is no improvement to be made in dioptric telescopes.

308.

This, however, is by no means the only inference to be drawn from the hypothesis of diverse refrangibility. Another immediate implication is that dioptric telescopes must be totally unusable, for anything on the objects which is white must appear to be completely coloured.

309.

Indeed, if the hypothesis were correct, not only dioptric telescopes, spectacles, and magnifying glasses but the whole visible world would have the appearance of being in total confusion. We see all the lights of the heavens through refraction; sun, moon, and stars, since they shine through a medium they are actually occupying, and astronomers are well aware that they have to take this into consideration, especially when observing their rising and setting. Why is it then, that when we see all these luminous images, these greater and lesser lights, that are not chromatically variegated — they are not dispersed into the seven colours?

310.

Newton certainly feels the force of this argument; for after spending some considerable time on measurements and calculations as a consequence of the above proposition he simply comes out with the significant observation that; "tis a wonder that telescopes represent objects so distinct as they do". He then carries on with his calculations and shows that the aberration originating in the form of the glass is about 6,500 times smaller than that deriving from the colouration. Consequently, he cannot refrain from asking how it is that: "If the Errors caused by the different Refrangibility of the rays be so very great, it comes to pass, that Objects appear through Telescopes so distinct as they do?" At this stage, the informed reader will have no difficulty in assessing the way in which he answers this question. Here, the way in which he also avoids the issue and the peculiar manner in which he goes about things, have an oddity all their own.

311.

Even if he had been on the right track, however, even if he had real-ised, as Descartes did before him, that an edge is necessary for the appearance of prismatic colour, he would still have been able to main-tain that this aberration cannot be neutralised, that the appearance associated with the edge cannot be eliminated, and would, moreover, have been justified in doing so. For his opponents, too, Rizetti[32] and others, were unable to get to grips with the matter because of them being obliged, once the appearance associated with the edge had been recognised as constant, to ascribe it exclusively to refraction. It was only the later discovery that the appearance of colour is not merely a general effect, but also presupposes a particular chemical property on the part of the medium that opened up an approach which was not at first appreciated, but which we are now well able to explore.

Sixteenth experiment

312.

Here, Newton busies himself with comparing the appearance of col-our as produced by a prism with that brought about by lenses, and proving by experiment that they exactly coincide. He decides to use the setup of his second experiment, in which he employed a lens to

---

[32] Giovanni Rizzetti (1675–1751). An amateur Italian physicist who declared he was unable to successfully repeat Newton's experiments — although he did observe the dispersion of light. He considered that when this occurred light was being "tortured" (was this the origin of Goethe's similar statement?). In 1727 Rizzetti published his findings in *Die Luminis Affectionibus*. Desaguliers said of it: "[Rizzetti}...in a most arrogant manner has insulted the greatest philosopher that this or any other age ever bred, triumphing in what he thinks are the mistakes of Sir Isaac Newton and his own discoveries", (See *Philosophical Transactions*, 1728). See also Goethe's entry on Rizzetti in his *History*.

project a red and blue image, lapped with black threads, on to a screen, instead of the dual colouring of this image, however, he now makes use either of a white printed page or of a white page covered with black lines, on to which he projects the prismatic spectrum, the object being that he should observe degrees of distinctness or indistinctness in the imaging beyond the lens.

313.

Since what there is to say about this matter has been said comprehensively enough in the course of our treatment of the second experiment, we shall now confine ourselves to commenting briefly on Newton's procedure. He wants to show that, also in the case of the prismatic colours, the more refrangible of them have their point of distinctness closer to the lens and the less refrangible further from it. When it looks as though he is about to start demonstrating this, he makes a show of his extreme scrupulousness and observes that certain parts of the prismatic image do not lend themselves to this experiment; the deepest violet is so dark that one is quite unable to perceive the letters or lines in the image; after he has dealt with this in some detail and begun to investigate the red he speaks, as if in passing, of a sensible red; he then observes that, at this end of the spectrum also, the colours are so dark that the letters and lines cannot be recognised, and that one is therefore obliged to operate with the middle of the image, where they can still be seen.

314.

Thinking back over all that has been dealt with above, one realises that this qualification nullifies the significance of the whole experiment. If there is a place in the violet in which the letters become illegible and also one in the red in which they disappear, it naturally follows that in this case the figures on the most refrangible coloured surface are disappearing at the same time as those on the least refrangible one; conversely, where they are visible, there must be degrees of their being so at one and the same time. Consequently, what has to be taken into consideration here is not a diverse refrangibility of colours, but the fact that the more or less distinct appearance of these characters depends entirely upon the degree of brightness or darkness exhibited by the background. In order not to advertise what he is up to, Newton avoids any clear statements; he speaks of sensible red, although it is actually the black letters which remain sensible in the brighter red. In the extreme depth and darkness of the spectrum, red is anything but sensible, for it can no longer illuminate a page and the letters printed there are no longer perceptible. Newton also expresses himself in this vague manner when dealing with violet and the other colours. Sometimes they are treated as abstractions, sometimes as lights which illuminate the books; in this experiment, however, they cannot be regarded as illuminating and shining on their own account, but have to be regarded solely as a brighter or darker background to the letters and threads.

315.

Consequently, this is not only a case of this experiment relating to the second and being invalidated by it, but of it also helping to invalidate the experiment to which it relates. Here we find Newton openly confessing that the perceptibility of the black characters ceases at both ends of the spectrum and, moreover, that it does so because of the encroachment of darkness; from which it follows that increasing brightness will be accompanied by a corresponding increase in the distinctness of these characters, regardless of the colour. We shall provide a further summary of all there is to be said on this when describing the apparatus.[33]

---

[33] Presumably in the promised Supplementary Section.

Proposition VIII. Problem II

To shorten telescopes

316.

Here, Newton gives an account of his catoptric telescope, an invention which even after the improvement of the dioptric telescope has retained its value and importance. We, for our part, are concerned only with colours, and we therefore have no comment to make.

Newtonian Opticks
First Book, Second Part

317.

In this second part, too, erroneous and captious experiments are grouped together in a way which is not only confused but also intentionally misleading. They can be classified as either polemical or didactic in purpose.

318.

The author begins polemically; for once he imagines he has proved incontestably that colours really are contained in light, he is obliged to call into question all previous reliance upon experience, the established conviction that a boundary is necessary for chromatic appearances involving refraction; and he is of the opinion that he has managed to do this by means of the first four experiments.

319.

He then continues didactically, insisting yet again that once homogeneous light has been produced it is immutable and that there are various degrees of refrangibility. He busies himself with this from the fifth to the eighth experiment. Later on, in the seventeenth, he qualifies and even invalidates what he has proved in the fifth.

320.

From the ninth to the fifteenth experiment, however, he concerned himself with eliciting and demonstrating something which must have been of prime importance to him. If he has extracted the colours from colourless light and white surfaces, if he has indeed split white light up into colours, the reinfusing of what has been extracted, the reproduction of pure corporeal white, must involve his also being able to re-assemble what has been separated.

321.

We, however, are sufficiently persuaded that colour arises not from a dividing up of light but from the intrusion of an external condition, which expresses itself in various empirical forms — such as dimness, a shadow or a boundary, and we are therefore quite prepared for Newton's having to behave somewhat strangely once he attempts to show that pure white light consists of the ways in which this condition gives rise to conditioned, dimmed, overshadowed, or shaded light, once he attempts to produce a bright white by mixing dark colours.

322.

In that he now wants to check his first calculations, as it were, to demonstrate that what he has produced by mere division has to be added together again to yield what he started with, he finds that he is completely obstructed by what he thought he had avoided, the third external condition; he is therefore obliged to deny the senses, sensuous impressions, common sense, common language, everything which engages a man, an observer, a thinker.

323.

We believe that in the Historical Section, under the heading of *Newton's Personality* we have provided an adequate explanation of the psychological and ethical factors involved in this attitude. Our task here is simply to carry out our polemical duties.

Proposition I. Theorem I

The phenomena of colours in refracted Light are not caused by new modifications of the Light variously impress'd, according to the various Terminations of the Light and Shadow

324.

Since we have shown in our *Outline* that no colours are formed by refraction without there being a boundary between light and darkness, those who have been convinced by our exposition that this relationship constitutes the truth of the matter, will want to know what Newton does in order to make what is true untrue. As we shall now see in detail, he behaves here as he did in the first case, where it was his intention to make what is untrue true.

First experiment

See Figure 4, Plate XIII

325.

For if the sun shine into a very dark Chamber through an oblong hole F.

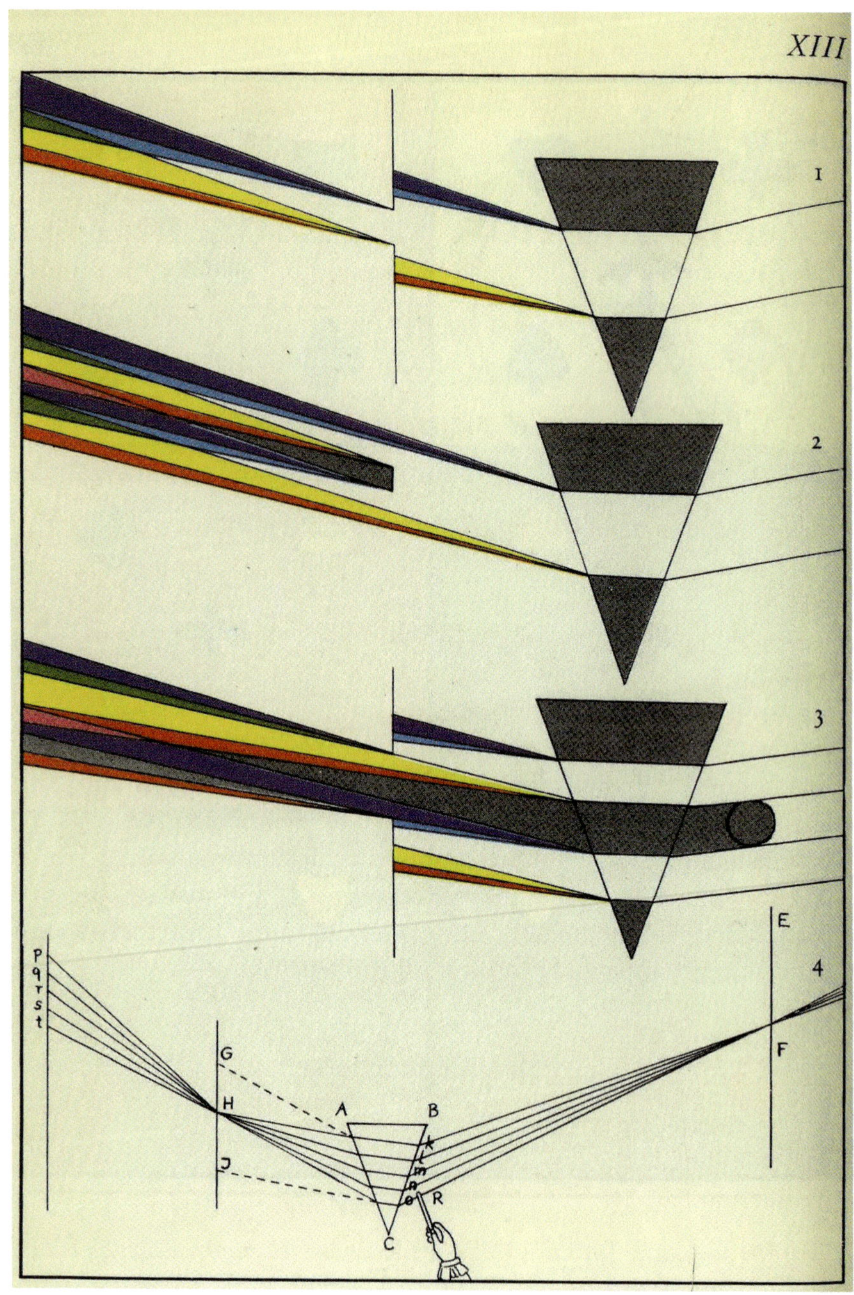

Plate XIII

326.

Although the hole must of necessity be elongated upwards, in the figure it is depicted only as a point and this makes it more difficult to grasp the situation.

327.

Whose breadth is the sixth or eighth part of an inch, or something less.

328.

So this initial apparatus consists of an extremely narrow slit, about six inches in height, in the shutterboard.

329.

And his beam FH do afterwards pass.

330.

Now it is already a beam again, although actually it is only an image of the sun, extremely narrowed in one plane and extremely elongated in the other.

331.

First through a very large Prism ABC, distant about 20 feet from the hole.

332.

Now, why 20 feet again? We have often had occasion to complain about this introduction of conditions without any indication being given of the reasons for them, and we have invariably found that they are either superfluous or arbitrary. This condition is arbitrary. What he actually wants is simply a very weak light, producing very weak colours, or perhaps his real objective is to make the experiment impossible. For who has at his disposal a darkened room 20 feet in length? And if anyone is so fortunate, how long does the sun at midday remain low enough to shine upon the wall opposite the window, or upon a prism which must of necessity be situated some distance above the level of the floor?

333.

We therefore declare this condition to be quite unnecessary, for the experiment is performed with a prism and no lens is involved — in which case the focusing and the imaging could necessitate operating at a particular distance.

334.

And parallel to it.

335.

That is to say, parallel to the table where the hole is to be found, parallel to the windowsill; as in all prismatic experiments, however, this means in fact that an imaginary line from the centre of the sun's image makes a right angle with the surface of the prism.

336.

And then (with its white part).

337.

Here once again, therefore, we have a white part of a beam which has already been refracted. It is, however, nothing more than the white centre of a very elongated image.

338.

Through an oblong hole H.

339.

This oblong hole is also depicted as a point, so that the illustration is completely misleading; the experiment requires that this hole, like the one in the shutterboard, should be oblong and vertical.

340.

Whose breadth is about the fourth or sixth part of an Inch.[34]

341.

That is to say, only a narrow slit. And why does this slit have to be so narrow? Simply so that one might not be able to see what is actually happening and being done.

342.

And which is made in a black opaque Body GI.

343.

It is not at all necessary that a metal sheet or pasteboard should be black, although it goes without saying that it should not be transparent.

344.

And placed at the distance of two or three Feet from the Prism.

345.

Once again, the distance is either irrelevant or incidental.

346.

In a parallel Situation both to the Prism and to the former hole.

347.

Since Newton presents his experiments not in a natural order but in an artificially arranged manner, he is obliged to describe the whole apparatus for every individual experiment. The same apparatus has

---

[34] *Vierten oder sechsten Teil eines Zolles.* This was incorrectly translated by Goethe. The original text was: "about the fortieth or sixtieth part of an inch". It is a requirement of the experiment that the slit should be as narrow as possible, and horizontal as well — not vertical, as stated by Goethe (who clearly could not understand the experiment).

often been used before, however, and had Newton adopted a straightforward procedure, he need only have referred back to the preceding experiment. In this case, however, each experiment is setup anew — that which is necessary being so infused with unnecessary conditions that a semi-darkness is created, and it is in this that he operates so readily.

348.

And if this white Light thus transmitted through the hole H, fall afterwards upon a white Paper p̲t̲, placed after that hole H, at the distance of three or four Feet from it, and there paint the usual Colours of the Prism, suppose red at I̲, yellow at s̲, green at I̲, blue at q̲, and violet at p̲.

349.

Take good note of this! The light is white when it arrives at the slit and forms the spectrum behind it. Now, pay particular attention to what follows.

350.

You may with an Iron Wire, or any such like slender opake Body, whose breadth is about the tenth part of an Inch, by intercepting the Rays at k̲, l̲, m̲, n̲, or o̲.

351.

Consider the figure and note where the rays at k̲, l̲, m̲, n̲, and o̲ are supposed to be. These letters are placed in front of the prism where it faces the sun, and as is evident from the five lines, they are therefore meant to represent coloured rays, where there is as yet no colour.[35] Up until now there has been no hint of anything like this anywhere

---

[35] No Newton's figure is correctly annotated; the letters are supposed to be placed by the leading edge of the prism. Goethe was not alone in misunderstanding this point. *See, for instance: Goethes Werke: Hg. im Auftrage der Grossherzogin Sophie von Sachsen,* pt. II, vol 2 (Weimar, 1890) p. 317; S. Gruner, 'Goethe's criticism of Newton's *Opticks', Physis,* 16, (1974), p. 76.

in the work, in any of the experiments, no suggestion that we are expected to accept and admit something which even runs counter to the author's own conception of the matter.

352.

What does the rod R do when it moves about in front of the prism? It severs the colourless image into several parts, makes several images out of one. The result is, of course, that the effect at p, q, r, s, and t is disturbed and contaminated; but Newton interprets the appearance as follows:

353.

Take away any one of the Colours at t, s, r, q, or p whilst the other Colours remain upon the Paper as before; or with an Obstacle something bigger you may take away any two, or three, or four Colours together, the rest remaining.

354.

The appearances to which this first experiment gives rise are depicted in accordance with the truth of the matter in the first three figures of our thirteenth plate. Since it is by describing and explaining this plate that we shall be developing the matter in more detail, we take the liberty of referring our readers to it, and begin by simply asking what it is that Newton has undertaken in order to prove his proposition.

355.

After maintaining that edges or boundaries between what is bright and what is dark have no influence upon the appearance of colour in refraction, what does he do in his experiment? He creates boundaries three times in order to prove that boundaries are of no significance.

356.

The first boundary is the top and bottom of the hole in the shutter F. He still has white light in the centre, but he does not reveal that colours are already in evidence at both ends. The second boundary is formed by the slit H. For, how could the refracted light, which is white when it reaches the screen GI, become coloured, if this were not a case of the boundary at the top and bottom of the slit H giving rise to prismatic colours? He now holds the third obstruction, a piece of wire, or any other such cylindrical body, in front of the prism, and thus creates further boundaries, an image within an image, with the colouring being the reverse of that at the edges of the rod. More precisely, a purple colour appears in the middle, with blue on one side and yellow on the other. He now imagines that with this rod he can take away coloured rays, whereas all he really does is project a narrow and fully coloured image on to the screen GI. He then goes on to operate with this image in the hole H, displacing and adulterating the colours that have formed there; what is more, he even hinders the formation of these colours, for it is in the hole H that they are first formed. To anyone who has come to understand the relationships involved here, this is certainly astonishing. Why should anyone indulge in so much covert manoeuvring in order to blur the significance of a phenomenon? Why, in this particular instance, should such a talented person do precisely what he usually rejects What ensues is, therefore, also completely at odds with experience.

357.

So that any one of the Colours as well as violet may become outmost in the Confine of the Shadow towards p̱, and any one of them as well as red may become outmost in the Confine of the Shadow towards ṯ.

358.

An inattentive observer could certainly be deceived by this, for with his obstacle ṟ he does create new colours in that old ones are displaced; but it can be said without hesitation that the way in which Newton gives expression to the matter is erroneous. Although he can certainly create confusion with the middle colours, he is unable to do so on the boundary; no green will ever be seen at p̱ or ṯ. Take good note of the following passage in which — like Bileam[36] — he begins once again to say the opposite of what he intends.

359.

And any one of them may also border upon the Shadow made within the Colours by the Obstacle R.

360.

So he now admits that he forms shadows with his obstacle R, that these shadows can be seen to have coloured fringes, and he takes this to be proof that the boundary between light and shadow does not run into colour! Is there any other instance in the history of the sciences to match this for obstinacy and effrontery?

361.

And, lastly, any one of them by being left alone, may border upon the Shadow on either hand.

362.

It is not denied that one can allow any already existent colour of the prismatic image to pass separately through an opening and thus isolate it; naturally, something similar can be brought about by making use of the rod; despite the shortcomings of such experimentation, however, the attentive observer will notice something which may well escape the inexperienced eye, namely that, even on this existent colour, it is the restriction which gives rise to the emergence of the contrasting colour. Newton thus futilely concludes that:

363.

All the Colours have themselves indifferently to any confine of shadow.

---

[36] Bileam — an Old Testament prophet, (Numbers 22–24). Described as a diviner, he was summoned by the Moabite King Balek to curse the Isrealites. But he was rebuked by his ass for considering such an evil thing, and concluded by blessing Israel.

364.

We have shown in detail in the *Outline* that in refraction the borders of the shadow act upon the colours in accordance with strict laws.

365.

And therefore the differences of these Colours from one another, do not arise from the different Confines of Shadow, whereby Light is variously modified, as has hitherto been the Opinion of Philosophers.

366.

Since his premises are false and his whole exposition is erroneous, his conclusion is also null and void. We hope to restore the honour of the ancient philosophers, for, although they did sometimes allow themselves to be sidetracked, the broad outlines of the pre-Newtonian approach to the phenomena were sound enough. We see more clearly what he was driving at by the way in which he concludes his exposition.

367.

In trying these things 'tis to be observed, that by how much the holes F and H are narrower, and the Intervals between them and the Prism greater, and the Chamber darker, by so much the better doth the Experiment succeed; provided the light be not so far diminished, but that the Colours at <u>pt</u> *be sufficiently visible.*

368.

So he admits that because of the distance from the window and the distance of the screens from the prism, the lights with which one is operating are extremely weak. Since the holes can scarcely be called slits, the coloured image has hardly any width, and yet one is supposed to be able to observe precisely which colour constitutes the actual boundary. The only real purpose of it all is to make the matter inaccessible to the senses, to produce colours which are so pale that one is better able to operate in them when using the rod R. As we shall now show, anyone who performs the experiment in a powerful light will find the falsehood of Newton's assertion obvious enough.

369.

To procure a Prism of solid Glass large enough for this Experiment will be difficult, and therefore a prismatick Vessel must be made of polish'd glass Plates cemented together, and filled with salt Water or clear Oil.

370.

We have criticised Newton before for repeating the description of his apparatus with every experiment, for not clarifying the relationship between the different experiments performed with the same apparatus. Similarly, it may also be observed here that Newton always uses his water prism when he requires a white centre and thus has to displace a large image by refraction.

371.

It is worth noting how, first, he slips this white centre in through the back door and then gradually allows it to take over until its bounding edges are no longer mentioned at all; and this all takes place before the very eyes of the learned and experimental world, which is usually so precise and inquisitive.

Second experiment

372.

Since this experiment belongs to the heterogeneous group in which prisms and lenses are used in combination it can only be covered properly in the Supplementary Section which we have often referred to. Moreover since Newton follows it with its exact equivalent — an experiment which he himself admits is not only easier to perform, but when carefully analysed renders its precursor quite superfluous — it can be disregarded all the more readily.

Third experiment

See Figure 1, Plate XIV

373.

Such another Experiment may be more easily tried as follows. Let a broad beam of the sun's Light.

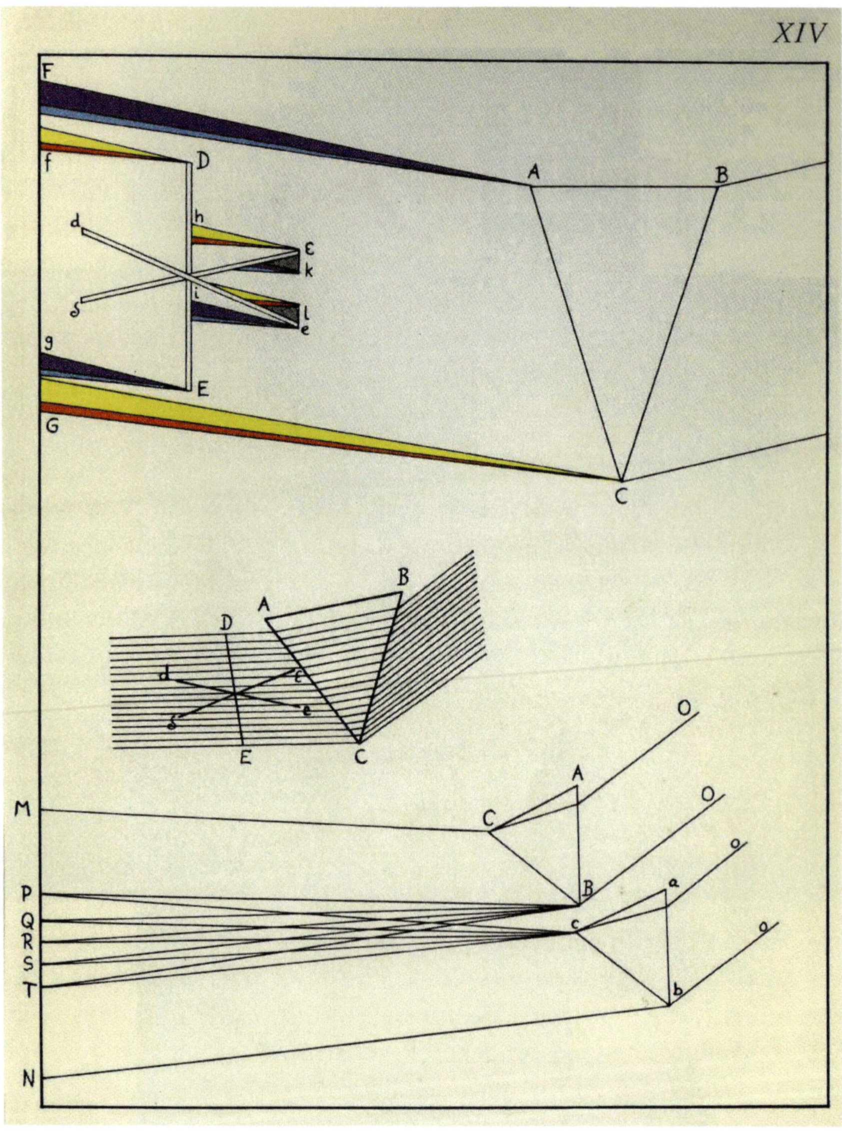

Plate XIV

374.

Now it is a wide sunbeam. But all this means is that the hole through which the light enters is enlarged; or rather that the prism is placed in the open sunlight — a situation unique to this experiment.

375.

Coming into a dark Chamber a hole in the Window shut be refracted by a large Prism ABC.

376.

Our ordinary water prism is admirably suited to this experiment.

377.

Whose refracting Angle C is more than 60°.

378.

This increase in the size of the angle is quite pointless, particularly so in this experiment, since it is a condition which makes it more difficult to perform a perfectly straightforward test by necessitating the construction of some extraordinary awkward apparatus.

379.

And so soon as it comes out of the Prism, let it fall upon the white Paper DE glewed upon a stiff Plane; and this Light, when the Paper is perpendicular to it, as 'tis represented in DE, will appear perfectly white upon the Paper.

380.

Now, at last, we have a light which has passed through a prism, been refracted, and yet is still completely white. Although our readers may find it tiresome, we must now draw their attention to how this has come about.

381.

Firstly, in the third experiment of Part One, we are introduced to a fully coloured spectrum, from which proof of diverse refrangibility is extracted by means of various experiments and deductions. Once this has been accomplished the author allows a light which is still white in spite of having been refracted, to appear unannounced at the end of the illustration of the fifth experiment. Soon afterwards he introduces an image which, though uninterruptedly coloured, also has a white centre. He then begins to operate within this white centre, some-times, moreover, without admitting its existence, and now, since he does not recognise the effect of the boundary between light and shade, he denies that there is any appearance of colour on the screen DE. Why is no mention made of the coloured fringes which emerge at both ends, A and C, of the back face of the prism?[37] Why should the screen DE not be any larger? The reason is, presumably, that if it were, one would have to take into account the fringes that would appear there.

382.

On consulting the figure, one sees a stream of lines approaching the prism, passing through it, and emerging on the other side; and this linear stream is supposed to represent a space which is white through-out. It is, moreover, by means of these fictitious lines that we are provided with yet another pictorial representation of the hypothetical beams. Now note what happens on the screen DE. When it is placed at de, what happens at <u>e</u>? The refracted light is still white when it reaches the edge of the screen, and at this edge immediately begins to form one set of colours which, in this position, consists of yellow and yellow-red. Because of the angle of the screen, the boundary colours which develop here spread over the whole screen to create the phenomenon he is describing; at precisely the point where Newton denies the existence of an edge or boundary he introduces an edge in

---

[37] Because they play no part in the experiment; see para. 395 for the explanation.

order to exhibit the phenomenon he is discussing. In the position δε the converse appearance is observed, namely, a violet edge; and this also spreads over the whole screen, as is clearly apparent from the faithful representation of the matter provided by our figure.

Since Newton could not admit that in this case it is the edge of the screen which has come into play, his rigid interpretation remains unrevised, and he continues as follows:

383.

And if the Light before it fall upon the Paper be twice refracted the same way by two parallel Prisms, these Colours will become the more conspicuous.

384.

So, a light can be refracted twice by two successive prisms and yet remain white and project as such on to the screen DE? Note this well! For it is to be expected that a twice refracted white light should produce colours which are more vivid, the displacement of the image being duplicated. Not only is it difficult to setup the required apparatus — Newton stipulating that two water prisms should be employed, each with a refracting angle in excess of 60° — but also, there is no point in recommending such an intensification of the experiment; the colours obtained by operating with a *single* prism are distinct enough, and anyone who does not see where they come from will gain nothing from the second prism. Nevertheless, Newton continues:

Here all the middle parts of the broad beam of white Light which fell upon the Paper, did without any Confine of Shadow to modify it, become colour'd all over with one uniform Colour.

386.

We have already pointed out that in this instance the edge of the cardboard itself constitutes the boundary and throws its own coloured half-shadow over the paper.

387.

The Colour being always the same in the middle of the Paper as at the edges.

388.

Certainly not! The careful observer will have no difficulty in picking out yellow-red from which yellow develops on one boundary, and blue from which violet radiates on the other.

389.

And this Colour changed according to the various Obliquity of the reflecting Paper, without any change in the Refractions or Shadow, or in the Light which fell upon the Paper.

390.

He tilts his cardboard backwards and forwards and yet maintains that the circumstances have remained the same. He also asserted this, and with just as little justification, in the previous experiment. Since he consistently overlooks the main points and never bothers about his assumptions there is no variation in his *therefore.*

391.

We are reminded here of Basedow[38] who was a heavy drinker, and who, when he was in his prime used to display a most agreeable brand of humour when in good company. He used to maintain that the conclusion *ergo bibamus,* could be drawn from any premise. The weather's fine, therefore let's drink to it! It's a dreadful day, therefore let's drink to it! We're all friends together, therefore let's drink to it! There are some horrible blighters amongst us, therefore let's drink to it! Newton also derives his *therefore* from the most varied premises.

---

[38]Johann Basedow (1723–1790) was a highly influential German educational reformer who advocated the use of realistic teaching methods and the introduction of Nature study, physical education, and manual training into schools.

The refracted light image is wholly uninterruptedly coloured; therefore light is diversely refrangible. It has a white centre, and yet it is still diversely refrangible. And after having made use of double and triple edges and boundaries of light and shadow in these three experiments, he draws his conclusion in precisely the same way:

392.

And therefore these Colours are to be derived from some other Cause than the new Modifications of Light by Refractions and Shadows.

393.

He has transmitted this sort of logic to his followers, who have continued to repeat their interminable *ergo bibamus* to the present day; it is just as laughable as Basedow's, but their constant reiteration of the same joke is much more irksome than his could ever have been.

394.

It goes without saying that the author is now ready to provide his own account of the cause, and he continues as follows.

395.

If it be asked, what then is their cause? I answer, that the Paper in the posture de. *being more oblique to the more refrangible Rays than to the less refrangible ones, is more strongly illuminated by the latter than by the former, and therefore the less refrangible Rays are predominant in the reflected Light.*

396.

Note the extraordinary about-turn he has to make in order to explain his phenomenon. He began with a refracted and yet wholly white light. One sees no colours in it when the screen is vertical; but, when it is tilted, they appear immediately. Since he is not interested in edges and borders, which only affect the sides, he assumes that when the screen is in a more oblique position the whole spectrum is actually present, but that only one end of it is visible. Why, then, is the green which borders on the yellow never visible? Why is it that when the yellow is played backwards and forwards over the white screen it always finishes up white? It is only natural that there should never be any appearance of green when this is done, for the yellow and the yellow-red fringe is only effective on one side and the other fringe is unable to meet it. This second fringe also exhibits its one-sided effectiveness; blue and violet form, without yellow and yellow-red being able to escape and spread out.

397.

In order to make it perfectly clear that these colours are formed solely by the edge, we devised a variation on this experiment involving a screen with raised areas on it, as well as tin-tacks and segments of a sphere. This provided convincing evidence that it is only the shadow-casting boundary within the refracted, but still white, light which is able to produce colours.

398.

Where-ever they are predominant in any Light, they tinge it with red or yellow, as may in some measure appear by the first Proposition of the first Part of this Book.

399.

This "in some measure" is Newton's way of saying "not at all". For it is only in so far as the proposition is proved that anything can appear or emerge from it; now we have shown in detail that it cannot be proved, and cannot, therefore, be adduced in confirmation of anything.

400.

And will more fully appear hereafter.

401.

We hope to have settled with the hereafter as well as the past.

Fourth experiment

402.

Here, Newton considers the case of soap bubbles, which change colour without one being able to say that there is any change in the boundary between light and shadow. This is quite the wrong place for such a consideration. We have made it quite clear in our *Outline* that the appearances on soap bubbles belong to a completely different field.

403.

Although it is a general assertion that light and shadow, light and non-light are required for the formation of colour, it can arise in a variety of ways. With refraction, it results from a particular action, namely, the displacement of a boundary between light and shadow.

404.

To these Experiments may be added the tenth Experiment of the first Part of this first Book.

405.

We can disregard what is said here, for we have already taken the context into consideration in our own interpretation of the experiment.

Proposition II. Theorem II

All homogeneal Light has its proper Colour answering to its Degree of Refrangibility, and that Colour cannot be changed by Reflections and Refractions

406.

In the Experiments of the fourth Proposition of the first Part of this first Book, when I had separated the heterogeneous Rays from one another.

407.

The extent to which this can be disregarded as a clear-cut separation has already been made clear to our readers, and Newton himself will soon make a further confession concerning the nature of this segregation.

408.

The Spectrum pt formed by the separated Rays, did in the Progress.

409.

So here we have progress! But is it steady progress?

410.

From its End p̲. on which the most refrangible Rays fell, unto its other End t̲, on which the least refrangible Rays fell, appear tinged with this series of Colours.

411.

Take good note of the "series".

412.

Violet, indigo, blue, green, yellow, orange, red, together.

413.

Take good note of "together".

414.

With all their intermediate Degrees.

415.

That is to say, although there were no gaps within the series, there were degrees. Now note what follows:

416.

In a continual Succession perpetually varying.

417.

We have just had separated colours and now have a continual succession of them; how surreptitiously falsehood and truth are intermingled here in what one might also call an uninterrupted succession: the falsehood that in this experiment the colours are separated, the truth is that they appear in an uninterrupted succession.

418.

So that there appeared as many Degrees of Colours, as there were sorts of rays differing in Refrangibility.

419.

Here, once again, we have degrees. In an uninterrupted series such as that which Newton has in mind there are no natural degrees, only artificial ones; nevertheless, although he denies Nature, our readers are already aware of part of the way in which his artificial structure of degrees is covertly supported by it, and we shall have to return to the other part later on.

Fifth experiment

420.

Now, that these Colours could not be changed by Refraction, I knew by refracting with a Prism sometimes one very little Part of this Light, sometimes another very little Part; for by this Refraction the Colour of the Light was never changed in the least.

421.

Since we have already indicated what happens here, we have only to take note of the immediate effect of these absolutist assertions concerning *never changed in the least.*

422.

We shall now anticipate an observation which really belongs to the history of colour theory. In Haüy's *Treatise on Natural Philosophy*[39] the above assertion is repeated in Newton's peremptory words: the German translator, however, is obliged to add the following footnote: "I shall subsequently take the opportunity to give an account of my own experiments, and indicate the kinds of light in the colour spectrum of which I have found this to be true and those of which I have found it to be untrue". Note that the tenability of the Newtonian doctrine depends solely and exclusively upon the absolute assertion of a tenet which is valid and, yet, invalid. Since

---

[39] René Just Haüy (1743–1832). A French mineralogist, and one of the founders of the science of crystallography.

Haüy propounds the Newtonian doctrine without qualification, it is without qualification drummed into the head of every young Frenchman who passes through the lyceums; the German is obliged to introduce qualifications, and yet this theory, the plausibility of which is incompatible with qualifications, is still the correct one; it is published, translated, and the public has to continue to pay through the nose for this fairy tale.

With respect to such qualifications, however, Newton, as we shall now see, gave his followers the lead in a truly exemplary manner.

423.

If any Part of the red Light was refracted, it remained totally of the same red Colour as before.

424.

Once again he starts out with the red which is so convenient to him, doubtless in order that every other experimenter may do so and become so involved in the problems it presents that he either fails to investigate the other colours, or is at least prejudiced in his view of the appearances. Hence the peremptory assurance with which the author now proceeds.

425.

No orange, no yellow, no green or blue, no other new Colour was produced by that Refraction. Neither did the Colour any ways change by repeated Refractions, but continued always the same red entirely as at first.

426.

What actually happens is fully explained above.

427.

The like Constancy and Immutability I found also in the blue, green, and other Colours.

428.

If the author has a clear conscience, why does he now mention the colours out of order? Why does he not mention yellow — on which the opposing edges show up so distinctly? Green also exhibits them quite unmistakably, and this is probably why he mentions it last.

429.

So also, if I looked through a Prism upon any Body illuminated with any part Of this homogeneal Light, as in the fourteenth Experiment of the first Part of this Book is described; I could not perceive any new Colour generated this way.

430.

What actually happens, we have already explained above.

431.

All Bodies illuminated with compound Light appear through Prisms confused, (as was said above) and tinged with various new Colours, but those illuminated with homogeneal Light appeared through Prisms neither less distinct, nor otherwise colour'd than when viewd with the naked Eyes.

432.

Only if one's eyes are extremely bad or one's mind is completely clouded by prejudice will one see or speak like this.

433.

Their Colours were not in the least changed by the Refraction of the interposed Prism.

434.

Bear in mind for a moment this unconditional *not in the least,* and note what follows!

435.

I speak here of a sensible Change of Colour.

436.

Something must, indeed, be sensible if it is to be observed.

437.

For the Light which I here call homogeneal.

438.

Here we have the hetman of the Cossacks again.

439.

Being not absolutely homogeneal, there ought to arise some little Change of Colour from its Heterogeneity. But if that Heterogeneity was so little as it might be made by the said Experiments of the fourth Proposition that Change was not sensible.

440.

If one refers back to what we have said about these experiments, in the course of which consideration was also to the passage now in question, it will be quite clear that there is no way in which the so-called Newtonian heterogeneity can be reduced, and that this is all a matter of sham fighting, which he makes use of for his own sophistical purposes. In short, everything which he first expresses in its absolute form in his propositions he subsequently qualifies, taking refuge either in the indefinite or in the discernible, as also in the present case, where he concludes:

441.

And therefore in Experiments, where Sense is Judge.

442.

Another characteristic expression. The senses are not judges at all; although, if they are healthy in their external relationships and uncorrupted within, they are excellent witnesses.

443.

That Heterogeneity ought to be accounted none at all.

444.

The snake is biting its own tail again, and for the hundredth time we witness precisely the same procedure. Although in the first instance the colours are totally unchangeable, a certain change is eventually perceived, this perception is then pestered into becoming increasingly insignificant, and although it remains apparent to the senses, is eventually declared to be completely devoid of importance. I would certainly like to know what physics would be like were every part of it to be dealt with in this manner.

Sixth experiment

445.

And as these Colours were not changeable by Refractions, so neither were they by Reflections. For all white, grey, red, yellow, green, blue, violet Bodies, as Paper, Ashes, red Lead, Orpiment, Indigo Bise, Gold, Silver, Copper, Grass, blue Flowers, Violets, Bubbles of Water tinged with various Colours, Peacock's Feathers, the Tincture of Lignum Nephriticum, and suchlike, in red homogeneal Light appeared totally red, in blue Light totally blue, in green, and so of other Colours.

446.

Were we not accustomed to finding Newton's assertions contradicted by experience, we might well have wondered how he could have asserted something so completely untrue. The experiment is such a simple one and is so easily performed that there is no difficulty in demonstrating to one and all that the statement is false.

This experiment really belongs to the chapter concerned with apparent mixing, and that is where we have dealt with it (O. 565, 566).

447.

Why does Newton pursue his end by employing coloured powders, flowers, and tiny bodies, which are difficult to handle? The experiment can be carried out much more easily, and anyone attempting to do the right thing will find it much clearer, if it is performed with larger coloured surfaces — coloured paper, for example, bringing out its implications most clearly.

448.

It goes without saying that all the colours of the image show up most clearly and vividly on a white surface. They also show up clearly on grey, but not so vividly as the grey shades into black there is a corresponding decrease in vividness. If coloured surfaces are employed, however, an apparent mixing takes place, and the colours of the spectrum appear — more vivid and beautiful in so far as they conform to the colour of the paper, more subdued and indistinct in so far as they contrast with it; in fact, in so far as they are able to blend with the colour of the paper, they form a third colour which actually becomes visible. Here, we have a true and natural relationship, which anyone can ascertain for himself by placing a prism in the sun and projecting the spectrum on to white, grey, or coloured paper.

449.

Note, in what follows, the author's characteristic procedure of going on to qualify what has already been stated.

450.

In the homogeneal Light of any Colour they all appeared totally of that same Colour, with this only Difference, that some of them reflected that Light more strongly, others more faintly.

451.

The appearance can only be said to be *strong* and *faint* in the case of white, grey, and black; in the case of all coloured surfaces, however, as has already been observed, it is the blending which has to be noted, for it is then that which has just been described takes place.

452.

I never yet found any Body, which by reflecting homogeneal Light could sensibly change its Colour.

453.

Here, once again, we have the word sensible: it is indeed most sensible that purple arises almost immediately when the yellow-red end of the spectrum is cast on to blue or violet paper, and the situation is the same with any other mixture we might care to employ. The element of captiousness in the way in which Newton performs the experiment with bodies or corporeal objects, powders, etc. should not be overlooked, however, for the light reaches the eye not from a plain surface, but from projections and cavities, illuminated and shaded places, and the experiment is therefore inconclusive and confused. We therefore insist that it should be performed with finely coloured papers stretched smoothly over cardboard. It will yield more or less satisfactory and clear results if use is made of taffeta, satin, or fine cloth. As might have been expected, Newton concludes with his *ergo bibamus*, rounding off the matter with a great flourish:

454.

From all which it is manifest, that if the sun's Light consisted of but one sort of Rays, there would be but one Colour in the whole World, nor would it be possible to produce any new Colour by Reflections and Refractions, and by consequence that the variety of Colours depends upon the Composition of Light.

455.

Those of our readers who know what the assumptions are worth will be able to evaluate the conclusion for themselves.

Definition

456.

The homogeneal Light and Rays which appear red, or rather make Objects appear so, I call Rubrifick or Red-making; those which make Objects appear yellow, green, blue, and violet, I call Yellow-making, Blue-making, and so of the rest. And if at any time I speak of Light and Rays as coloured or endued with Colours, would be understood to speak of Light and Rays as colour'd or embued with Colours, I would be understood to speak not philosophically and properly, but grossly and accordingly to such Conceptions as vulgar People in seeing all these Experiments would be apt to frame. For the Rays to speak properly are not coloured. In them there is nothing else than a certain Power and Disposition to stir up a Sensation of this or that Colour. For as Sound in a Bell or musical String, or other sounding Body, is nothing but a trembling Motion, and in the Air nothing but that Motion under the Form of Sound; so Colours in the Object are nothing but their Dispositions to propagate this or that Motion into the Sensorium and in the Sensorium they are sensations of those Motions under the Forms of Colours.

457.

Our first task is to find some sort of explanation for the way in which this curious piece of theorising is inserted here under the rubric of a definition, for only by so doing can we improve our understanding of the matter. We learn from the history of colour theory that, as soon as Newton presented his explanation of the prismatic phenomenon, contemporary physicists saw very clearly that this understanding of the matter implied that colours are corporeally present in light, confronted him with the then extremely popular wave theory, and maintained that with it one could explain or conceive of the colours with greater ease and effectiveness. Newton countered that it was all the same what theory was employed to explain these phenomena; all that mattered to him was the fact that these colour-forming properties of light are manifested during refraction, reflection, inflection etc. Since this wave theory, this likening of colour to sound, was taken up a new by Malebranche[40] it also found favour in France. This definition or declaration is therefore inserted here in order to do away with and neutralise this theoretical disagreement, in order to amalgamate the atomism of the Newtonian concept with the dynamism of his antagonists, so that it really does seem as if there is no difference between the two theories. The reader may care to judge the matter for himself, noting the interweaving of dynamic and corpuscular terms.

458.

Our account gives the answer to those who raise the question of how could the Newtonian theory of colours still enjoy general acceptance after Euler's subsequent stimulation of interest in the wave theory. The point is that, although it was realised that various vibrational properties are concealed in light, and that they are brought out by refraction and other external conditions, no further progress was made; as Newton himself pointed out during this controversy.

459.

Note now, however, how Newton makes use of this state of affairs. He plans to measure the relative proportions of the colours of his spectrum, and to compare these proportions with those of sound; the wave theory having its uses for this purpose.

---

[40] Nicholas Malebranche (1638–1715) was a Roman Catholic priest, theologian, and a major philosopher of Cartesianism who was the first to clearly enunciate (in 1699) the principle that colours depend upon the frequency or period of vibrations.

Proposition III. Problem I

To define the Refrangibility of the several sorts of homogeneal Lights answering to the several Colours

Seventh experiment

460.

The author, who may well have thought that his theory of colours was only known to physicists, that his interpretation of the phenomena could not easily be related to descriptions of other natural appearances, now sets about relating the dimensions of his spectrum to those of sound, and by means of this to impart a degree of credibility to his view.

461.

Thus, he quite futilely links the present experiment with the fifth one of Part I and with the one related to the fourth proposition; he takes his normal spectrum, allows it to fall upon some paper upon which an outline is drawn, and then draws transverse lines at the boundaries between the colours, in order to measure the area which each occupies, and the distances between them.[41]

---

[41] See Figure 4 in the Appendix. Goethe believed that there are only six colours in the visible spectrum (see his colour circle illustrated on the front cover). On this occasion,

462.

Having wasted much time and paper attempting to prove, contrary to Nature, that the spectrum is made up of an infinite array of overlapping coloured discs, he then suddenly draws transverse lines through the boundaries where they encroach upon each other, where each may be differentiated from the other.

463.

It is all the more difficult to judge the author's work because it is such an admixture of truth and error, here for example, he introduces the truth that the boundaries between the colours in the perpendicular spectrum may be presented fairly well by horizontal lines; this truth, however, is accompanied with the falsehood, backed up and given an appearance of verity by measurements and calculations, that the relative proportions of these are fixed.

464.

Our readers are sufficiently aware of the situation with respect to these two points. If they require a concise revision of it they only have to refer to our fifth plate. There we have depicted the displacement of a bright image in a series of rectangles, which is the most convenient way of presenting what the situation involves. The colours of the illustrated cross-sections appear between horizontal parallel lines. At first they are separated by white, then yellow and blue superimpose, giving rise to green. The latter finally predominates, for the yellow and blue merge into it. This plate shows clearly that each

---

it was probably Newton who was being fanciful, by adding indigo; in order to make out that there are seven colours (see the prismatic spectrum illustrated on the back cover). The motivation for this was undoubtedly his attempt to support the notion that there might be a correlation between sound and colour; more specifically between the diatonic scale and the visible spectrum. Accordingly, he divided the various coloured bands into proportions commensurate with this notion. This was a view shared by many - from Aristotle to Scriabin (with his colour organ). See G. Biernson, "Why did Newton See Indigo in the Spectrum?" *American Journal of Physics*, 40, 526–533 (1972).

cross-section made through the advancing phenomenon is different, and that only the one with the dotted oval above it approximates fairly well to the Newtonian spectrum. The matter is made perfectly clear in the sixth plate, the situation with respect to a displaced dark image being the same.

465.

There would appear to be no point in our concerning ourselves with Newton's measurements, or with the figures and calculations he bases upon them, for we can look around his grots and follies whenever we please. It should be emphasised, however, that the thirds, fourths, and fifths on which he dwells are entirely imaginary, and provide no basis for comparing colour and sound.

Eighth experiment

466.

In the previous experiment the author claims to have measured the spectrum produced by the glass prism and then went on to miscalculate its proportions. He now proceeds to investigate combinations of various media in order to define the variegated appearance of colour in accordance with the law that has been discovered.

467.

To this end, he takes a water prism with its refracting angle pointing downwards, places in it a glass prism with its refracting angle pointing upwards, and allows the sunlight to pass through. He then experiments until he finds a glass prism which has a smaller angle than that of the water prism, but which by stronger refraction compensates for the refraction of the latter in such a way that the incident and emergent beams are parallel: it is the corrected refraction which is then supposed to account for the disappearance of the colour phenomenon.

468.

Since everyone recognises this experiment to be unacceptable, we shall disregard it and not make an issue of it; as soon as the achromatism of continual refraction or, conversely, the chromatism of corrected refraction had been discovered, it could hardly be denied that there was an important factor which Newton had failed to take into consideration.

469.

It was, however, quite understandable that Newton should not have investigated the matter very closely. Since he saw refraction itself as the cause of the appearance of colour, and had spoken of and defined refrangibility, diverse refrangibility, it was only natural that he should have equated effect with cause, that he should have believed and maintained that a medium which refracts more must also produce stronger colours; and that, in that it corrects the refraction of another medium, it must also give rise to an immediate removal of the appearance of colour. Since refrangibility stems from refraction, it must surely keep pace with it.

470.

It has caused surprise that someone with Newton's reputation for meticulous experimentation, that such a first-rate observer, should have been capable of carrying out such an experiment and of overlooking its main significance. Yet, Newton did not find it easy to carry out an experiment unless it favoured his opinion; or at least, he only perseveres with those which appear to confirm his hypothesis. And how was he to conceive of a diverse refrangibility which was itself different again from refraction? We shall enquire more fully into this in the *History of the Theory of Colours* where Dollond's[42] invention is discussed, the natural relationship having already been clarified in our *Outline* (O. 682–687).

---

[42] John Dollond (1706–1761), a maker of optical and achromatic instruments, devised an achromatic lens system by combining flint and crown glasses.

471.

The truth of the matter was that the discovery of achromatism dealt a death blow to the Newtonian doctrine. Although there were people, such as our own Klügel,[43] clear-headed enough to sense that this was so, what they actually said tended to be somewhat equivocal. The school, however, which had already had plenty of practice in theory patching, in pasting and plastering, had no difficulty in producing surgeons to embalm the corpse. Hence the truly Egyptian manner in which the theory has presided posthumously over gatherings of natural philosophers.

472.

One spoke not only of "diverse refrangibility" but also of a "diverse dispersability", thereby endowing the vague word dispersion, which had already been employed by Grimaldi,[44] Rizzetti, Newton himself, and others, with a wholly peculiar significance. Although it was an inapposite term, it was applied to the newly discovered phenomenon, credited with immense significance and saved a theory by turns of phrase — a theory, moreover, which consisted of nothing more than turns of phrase.

473.

We shall bypass the measurements and calculations, which the physicists and mathematicians have already declared to be erroneous, and proceed to interpret and elucidate the concluding paragraph. This sets the scene for further gerrymandering, the object of which is to leave us in the dark, not enlighten us, for the author continues as follows:

---

[43] Georg Simon Klügel (1739–1812), was Professor of Mathematics and Physics at Helmstedt and Halle. He translated Priestley's *History of Optics* (London, 1772) into German in 1775.

[44] Franciscus Maria Grimaldi (1618–1663), a Jesuit priest from Bologna who also experimented with light and darkened rooms. He obtained colours and dispersion of a sunbeam; but, by using two narrow slits instead of a prism. It was quite a different effect, quite a different type of dispersion from that produced by Newton with his prism; and much more delicate.

474.

And these Theorems being admitted into Opticks.

475.

It really is astonishing that it should be precisely in a context which is generally recognised as being erroneous that he should make this suggestion.

476.

There, would be scope enough of handling that Science voluminously after a new manner, not only by teaching those things which tend to the perfection of Vision, but also by determining mathematically all kinds of Phenomena of Colours which could be produced by Refractions.

477.

It was precisely this that was done, on the model of the last experiment. The result was that for a long time it was impossible to perfect the dioptrical telescope or gain any real insight into the general nature of colour and particularly difficult to understand colour which stems from refraction. Now we have a simple transition to what comes next.

478.

For to do this, there is nothing else requisite than to find out the Separations of heterogeneous Rays.

479.

The extraordinary methods he adopted in attempting to achieve this and the paucity of what he achieved have already been surveyed in exhaustive detail. But note carefully what else he requires.

480.

And their various Mixtures and Proportions in every Mixture.

481.

So, first they have to be separated, and then they have to be mixed again, one has to establish their proportions both when separated and in the mixture. And to what end? It will soon become evident what the author has in mind, for it is by using these mixtures that he intends to put pressure upon us. For the time being he proceeds to feed us with extravagant promises.

482.

By this way of arguing I invented almost all the Phenomena described in these books.

483.

He has indeed invented them, or rather adapted them to his argument.

484.

Beside some Others less necessary to the Argument; and by successes I met with in the trials, I dare promise, that to him who shall argue truly, and then try all things with good Glasses and sufficient Circumspection, the expected event will not be wanting.

485.

The expected success will simply be that which has already been achieved — a hypothesis will be further elaborated and the preconceived idea established ever more rigidly within the mind.

486.

But he is first to know what Colours will arise from any others mix'd in any assigned Proportion.

487.

And thus would the author lead us gently to another threshold, politely pressuring us to pass over into another of his illusory concamerations.

Proposition IV. Theorem III

Colours may be produced by Composition which shall be like to the Colours of homogeneal Light as to the Appearance of Colour, but not as to the Immutability of Colour and Constitution of Light. And those Colours by how much they are more compounded by so much are they less full and intense, and by too much Composition they may be diluted and weaken'd till they cease, and the Mixture becomes white or grey. They may also be Colours produced by Composition, which are not fully like any of the Colours of homogeneal Light.

488.

Our primary task now is to attempt to make clear to our readers what this proposition is supposed to mean, how it actually relates to what precedes it and what it is intended to lead on to. By claiming that he has discovered within the spectrum a series akin to that of the musical scale, Newton has just given further support to the erroneous view that it basically consists of a fixed series of colours.

489.

We now know, however, that if one is to get to the bottom of the appearance, one has to envisage the displacement of a bright image taking place together with that of a dark one. Now, there are two colours, yellow and blue, which may be regarded as simple, two which are augmented, yellow-red and blue-red, and two which are mixed, green and purple. Newton was unaware of these differences, the only colours he examined being those created by the marked displacement of a bright image. It was these that he distinguished, counted, assumed to be five or seven in number, and even allowed to be innumerable, since any number of incisions may be made in a continuous series. And it is all these, regardless of how many of them there are, which are supposed to be the primitive, primary colours, the basic components of light.

490.

On closer examination, however, he was bound to find that many of these simple primaries looked exactly like other colours, which could be formed by mixing. If the exposition had been concerned with Nature itself, it would indeed have been difficult to explain how what is mixed could resemble or even be the same as what is primary and *vice versa*; the Newtonian approach makes this possible, however, and leaving aside general observations we shall therefore pass straight on to a discussion of the author's text. In calling our readers' attention to the final significance of this mixing and remixing, we shall make use of the usual medium of succinct observation.

491.

For a Mixture of homogeneal red and yellow compounds an orange, like in appearance of Colour to that orange which in the series of unmixed prismatick colours lies between them; but the Light of one orange is homogeneal as to Refrangibility, and that of the other is heterogeneal. And the colour of the one, if viewed through a Prism, remains unchanged, that of the other is changed and resolved into its component Colours, red and yellow.

492.

Since the author has inflicted such a variety of complicated experiments upon us, why does he not provide us with a precise description of this one? Why does he not refer back to some earlier experiment which might have helped us in some way? It probably resembles the one we introduced earlier (154 and 155), in which a couple of prismatic images are objectively superimposed upon each other, either wholly or in part, and then subjectively shifted apart by being viewed through a prism. Newton's only intention, however, is to prepare a way out for himself when repeated displacement of his homogeneous coloured images is accompanied by the appearance of new colours. What he wants to be able to say is that these images were not in fact homogeneous; and there is indeed no way of getting around anyone who expounds and argues in this manner.

493.

And after the same manner other neighbouring homogeneal Colours may compound new Colours, like the intermediate homogeneal ones, as yellow and green.

494.

Note the deftness of the author's approach! He selects his homogeneous green since green is generally recognised as a compound colour.

495.

Yellow and green therefore compound the Colour between them both.

496

In Nature and in our language this is something resembling a parrot green, formed by adding more yellow and less blue. But note what follows!

497.

And afterwards, if blue be added, there will be made a green the middle colour of the three which enter the Composition.

498.

So he begins by treating green as a primary colour and not recognising the yellow and blue from which it is compounded; then, after endowing it with a preponderance of yellow, he proceeds to remove this preponderance by mixing in blue, that is to say, he simply adds a further portion of fresh green and thus doubles the amount of green he began with. But he knows how to interpret the matter in an entirely different way.

499.

For the yellow and blue on either hand, if they are equal in quantity they draw the intermediate green equally towards themselves in Composition, and so keep it as it were in Equilibrium that it verge not more to the yellow on the one hand, and to the blue on the other, but by their mix'd Actions remain still a middle Colour.

500.

How much more easily he could have extricated himself had he given Nature its due and interpreted the phenomenon as it is — admitted that the prismatic blue and yellow, which are initially separated in the spectrum, subsequently combine to form a green, and that the green of the spectrum cannot be regarded as a primary colour. But this was out of the question! He and his school prefer words to the matter itself.

501.

To this mix'd green there may be farther added some red and violet. And yet the green will not presently cease, but only grow less full and vivid, and by increasing the red and violet, it will grow more and more dilute, until by the prevalence of the added colours it be overcome and turned into whiteness, or some other Colour.

502.

Here, again we have the prime fault of the Newtonian doctrine, its failure to take into consideration the *σχιερόν* of colour, its uncritical assumption that it is dealing only with lights. It is not lights which are being considered, however, but half-lights, half-shadows, which under certain conditions appear many coloured. If these half-lights and half-shadows are superimposed on each other there will certainly be a gradual relinquishing of their specific character in that they will cease to be blue, yellow or red, but in no way is this a matter of their being diluted. They are projected on to a patch of white paper and the paper is therefore darkened; a half-light or half-shadow is formed, consisting of many other half-lights or half-shadows.

503.

So if the Colour of any homogeneal Light, the sun's white Light composed of all sorts of Rays be added, that Colour will not vanish or change its Species, but be diluted, and by adding more and more white it will be diluted more and more perpetually.

504.

If the spectrum is allowed to fall upon a white screen standing in the sunlight it will appear as pale as would any other shadow on to which the sunlight falls without completely extinguishing it.

505.

Lastly, if red and violet be mingled, there will be generated according to their various Proportions various Purples, such as are not like in appearance to the Colour of any homogeneal Light.

506.

Here, at last we have purple, genuine, true, pure red, which verges neither upon yellow nor upon blue. The noblest of colours, the origin of which in physiological, physical, and chemical contexts has been adequately surveyed in our *Outline*, is completely absent from Newton's spectrum, as he himself admits, and this is surely due to his basing his observations exclusively upon the spectrum produced by the displacement of a bright image, to his not also considering that produced by the displacement of a dark image, to his drawing no parallel between the two. For, just as the displacement of a bright image eventually leads to a meeting of yellow and blue in the middle, so the displacement of a dark image eventually leads to a meeting of yellow-red and blue-red. The colour at the end of Newton's colour scale, which he calls red, is in fact merely yellow-red, so that there is no full red amongst his primary colours. But this is what happens with anyone who deviates from Nature, places what is first last, raises what is derived to the status of originality, debases what is original to the state of being derived, calls what is complex simple and what is simple complex. All who do so necessarily get everything wrong, since they began with what was wrong; and yet there are first-rate minds who also derive satisfaction from such error.

507.

And of these Purples mix'd with yellow and blue may be made other new Colours.

508.

It was, therefore, in this highly confused manner that the mixing and mingling was to be carried out; what he actually had in mind becomes apparent in what follows.

The colours were mixed in order to bring about the final neutralisation of their specific effect. His overriding purpose was to get them to produce white, and although in practice he was certainly unable to do so, he was constantly claiming that this was possible and feasible.

Proposition V. Theorem IV

Whiteness and all grey Colours between white and black, may be compounded of Colours, and the whiteness of the sun's Light is compounded of all the primary Colours mix'd in a due Proportion

Ninth experiment

509.

We have given the first half of this proposition adequate treatment in the chapter on real and apparent mixing, and our readers are able to assess the second half for themselves. Nevertheless, it is worth investigating how he attempts to prove what he has asserted.

510.

The sun shining into a dark Chamber through a little round hole in the Window shut, and his light being there refracted by a Prism to cast his coloured Image upon the opposite Wall: I held a white Paper to that image in such manner that it might be illuminated by the colour'd Light reflected from thence, and yet not intercept any part of that light in its passage from the Prism to the Spectrum. And I found that when the paper was held nearer to any Colour than to the rest, it appeared of that Colour to which it approached nearest; but when it was equally or almost

*equally distant from all the colours, so that it might be equally illumi-
nated by them all it appeared white.*[45]

511.

Consider what is involved in this operation. There is an incomplete
reflection of a coloured semi-bright image, which nevertheless obeys the
laws of apparent transmission (0. 588–592). We shall hear the author
out, however, and then place the true relationship in its context.

512.

*And in this last situation of the paper, if some Colours were intercepted,
the Paper lost its white Colour, and appeared of the Colour of the rest of
the light which was not intercepted, so then the Paper was illuminated
with Lights of various Colours, namely, red, yellow, green, blue, and violet, and every part of the Light retained its proper Colour, until it was
incident on the paper, and became reflected thence to the eye: so that if it
had been either alone (the rest of the Light being intercepted) or if it had
abounded most, had been predominant in the Light reflected from the
Paper, it would have tinged the Paper with its own Colour; and yet being
mixed with the rest of the Colours in a due proportion, it made the Paper
look white, and therefore by a composition with the rest produced that
Colour. The several parts of the coloured light reflected from the Spectrum,
whilst they are propagated from thence through the air, do perpetually
retain their proper Colours, because wherever they fall upon the eyes of any
spectator, they make the several parts of the Spectrum to appear under
their proper Colours. They retain therefore their proper Colours when they
fall upon the Paper, and so by the confusion and perfect mixture of those
Colours compound the whiteness of the Light reflected from thence.*

513.

As has been noted, the whole appearance is nothing but an incomplete reflection, it is essential to bear in mind that the spectrum itself

---

[45] See Figure 5 in the Appendix.

is a dark image made up of shadowy lights. The surface of the paper is white, but it is also rough, so that when the spectrum is brought up close to it, each of the colours will reflect from it in such a way that, although the reflections are weak, the careful observer will have no difficulty in distinguishing between them. Every point of the paper reflects all the colours simultaneously, however, so that they neutralise one another to a certain extent, and a kind of dimmed light results — to which no actual colour can be ascribed, the brightness of this dimmed light is like the dimness of the spectrum itself, but it is quite unlike the brightness of white light prior to its taking up and suffusing itself with colours. This is the cardinal point, and Newton constantly avoids it. For, although a grey can certainly be compounded from extremely bright colours, even material ones, there is never any chance of confusing it with white chalk, for instance, this is all so simple and straightforward in Nature itself: it has only been made so difficult and so infinitely complicated by these erroneous theories and sophistical reasonings.

514.

The truth of the matter is that, although that which is specific about the various colours is certainly obliterated by their combination, one cannot eradicate the σχιερόυ they have in common. One will become even more convinced that this is indeed the true relationship if one consults the way in which the matter has been treated in our chapters on apparent mixing and transmission, and then performs the experiment by allowing strong sunlight to fall upon coloured papers, from which it is reflected on to a surface standing in the dark.

515.

In the next three experiments, although Newton reveals more of his little contrivances and refinements, he leaves us in the dark with regard to the true nature of his apparatus and of the appearance it gives rise to. As is his wont, he presents the three experiments in the wrong order, beginning with the most complicated, continuing with one which is rather out of place, and ending with the simplest. In order to make things easier, both for ourselves and for our readers, we shall therefore reverse the order and begin with the:

Twelfth experiment

516.

The sun shining through a large Prism falls on a white Paper (and forms there a Whiteness).[46]

517.

Thus, Newton again works with light which, although refracted, is still uncoloured.

518.

A comb is placed immediately behind the Prism.[47]

519.

Note the incongruity of Newton's wanting to combine his white light with a card and a comb.

520.

The Breadths of the Teeth were equal to their interstices, and (the) seven Teeth.[48]

521.

As if one had been prepared for each major light beam.

522.

Together with their Interstices took up an Inch in Breadth. Now when the Paper was about two or three Inches distant from the Comb, the Light

---

[46] Goethe has misquoted Newton, who made no mention of a whiteness being formed.
[47] See Figure 9 in the Appendix.
[48] This should read "and seven teeth", not "and the seven teeth".

which passed through its several Interstices painted so many Ranges of Colours.

523.

Why does he not call them, "the series of prismatic colours"?

524.

Which were parallel to one another, and contiguous, and without any Mixture of white.

525.

And this phenomenon only occurred because each tooth had two edges, and because on reaching these edges the refracted and uncoloured light was immediately determined by them — a situation which is emphatically denied by Newton in the first proposition of this book. This is the extraordinary thing about his treatise: in the first instance the true relationships and appearances are rejected, then, if they can be used for some sort of purpose, they are immediately introduced as a matter of course, and without comment.

526.

And these Ranges of Colours, if the Comb was moved continually up and down with a reciprocal Motion, ascended and descended on the Paper.

527.

In no respect the same ranges of colours; there was a continual change in the appearances of colour at the edges of the moving comb, the images were constantly developing.

528.

And when the Motion of the Comb was so quick, that the Colours could not be distinguished from one another, the whole Paper by their Confusion and Mixture in the Sensorium appeared white.

529.

The expertise of our natural philosopher now enables him to so comb together his homogeneous lights that, once again, they provide him with a white which, once again, we are obliged to spoil. We have devised some apparatus which is very effective in making the relationships involved in this experiment fully apparent, the operation of moving a comb up and down at great speed is awkward and difficult. We therefore constructed a wheel of fine spokes which can be attached to the roller of a fly-wheel. The spoked wheel is placed between the large illuminated prism and the white screen. When it is set slowly in motion, each spoke in passing across the white region of the refracted image gives rise to a coloured strip in the familiar sequence; blue, purple, and yellow. These coloured appearances occur as each spoke passes, and succeed each other with ever greater rapidity as the speed of the wheel is increased. If the wheel is then accelerated to full speed, so that the observer is no longer able to distinguish between the spokes — the eye only perceiving a round disc — something remarkable happens: the image, which is white as it leaves the prism and which is coloured along its edges, appears quite clearly on the disc and, at the same time — since the latter can also be regarded as semi-transparent — projects on to the white screen beyond. Here, in an experiment which everyone will take pleasure in viewing, one has the true relationship before one's very eyes. There is no question of carding, felting, and combing ready-made coloured lights, the main point is the rapidity with which the apparent disc intercepts the whole image whilst also letting it pass through on to the white screen, where precisely because of the rapidity with which the spokes are moving, we are unable to perceive any colour at all. Although the centre of the image on the white screen is indeed white, it is dimmer and duskier due to it having been smothered and tempered by the disc, which has to be regarded as semi-transparent.

530.

The experiment is even more impressive if the coloured phenomenon is formed with a smaller prism, so that a complete spectrum is projected upon the spokes of the revolving wheel. If the apparent disc is spun rapidly the spectrum plays upon it in its full force, and in an equally undisturbed and unchanged form on the white screen behind it. How is the lack of mixing and confusion to be explained? Why is it that when the spoked wheel is fully accelerated, the ready-made colours are not whirled together? Why is it that on this occasion Newton chooses not to operate with his ready-made colours? Why does he posit developing ones? Merely because this enables him to maintain that they were indeed fully formed, and that it is by mixing that they have been changed into white; but the fact is that the space before our eyes remains quite simply because the passing spokes do not bring out the boundary which could give rise to colour.

531.

Since the author makes good use of his comb, he puts together quite a number of added experiments; he does not number them however, and we shall also deal summarily with their content.

532.

Let the Comb now rest, and let the Paper be removed further from the Prism and the several Ranges of Colours will be dilated and expanded into one another more and more, and by mixing their Colours will dilute one another, and at length, they will so far dilute one another, as to become white.

533.

The best way of determining what happens when black and white stripes alternate on the screen is to experiment subjectively. The edges form in the normal way on the boundaries of both the black and the white, the borders spreading over both the white and the black until the yellow border reaches the blue edge to form green, and the violet edge the yellow-red to form purple. Thus, we are able to simultaneously observe the displaced white and black images. If one stands further back from the cardboard the edges and borders intermesh to such an extent, unite so intimately, that one no longer sees anything but overlapping green and purple stripes.

534.

The same appearance can be produced objectively by operating with a comb in front of a large prism; one then has a good view of the shifting purple and green stripes on the white screen.

535.

Newton is therefore quite wrong when he suggests that all the colours intermesh, for only the colours of the opposing edges can mix, and it is precisely because of this that the other colours are kept apart. Since these colours are basically simply half shadows, they appear to be more rarefied when one steps back from the screen because they do not spread any further, because their effect weakens and gradually diminishes until it almost ceases, because each by itself becomes imperceptible, not because they mix to produce white. This experiment differs from certain others in that it cannot be said to involve neutralisation.

536.

With any Obstacle.

537.

Here, again is an obstacle, such as he used unsatisfactorily in the first experiment of this second part, and which he makes no better use of here.

538.

Let all the Light be now stopp'd which passes through any one Interval of the Teeth, so that the Range of Colours which comes from thence may be taken away, and you will see the light of the rest of the Ranges to be expanded into the place of the Range taken away, and there to be coloured.

539.

This is in no way the truth of the situation — a careful observer will see something quite different. If one covers up a gap in a comb, all one gets is a broader tooth which, if the gaps and the teeth are of equal breadth, is three times broader than a normal one. What happens on the boundaries of this broader tooth is no different from what happens on the boundaries of the narrower ones: the violet border extends inwards, the yellow-red edge marks the other side. Now it is possible that, at a certain distance, these two colours do not yet meet across the broad tooth although they have already done so across the narrower ones; in the first case, therefore one will see the yellow-red and the blue-red separated, whilst in the other cases purple will already be visible.

540.

Let the intercepted Range pass on as before, and its colours falling upon the Colours of the other Ranges, and mixing with them, will restore the Whiteness.

541.

Certainly not. As has already been indicated, at such a distance the series of colours passing through the narrow gaps of the comb simply become imperceptible, the result being an equivocal, which tends to be coloured rather than colourless.

542.

Let the Paper be now very much inclined to the Rays so that the most refrangible Rays may be more copiously reflected than the rest, and the white Colour of the Paper through the Excess of those Rays will be changed into blue and violet. Let the Paper be as much inclined the contrary way, that the least refrangible Rays may be now more copiously reflected than the rest, and by their Excess the Whiteness will be changed into yellow and red.

543.

It is quite evident that now Newton is simply raising his stakes on the third experiment of the second part.

Since this gaming expression has come to mind, one might go on to compare him to a card sharp who takes advantage of an inattentive banker to treble his stakes on a card he has not won, and who by luck and craft goes on to increase its value by turning down one comer after another. Whereas there he was operating in white light, here he is operating again in a light which has passed through a comb, and at a distance at which the colouring effects of the teeth are considerably weakened. But this is still a refracted light, and by placing the obstacle close to the screen one can reform the shadows and coloured borders. Therefore, the third experiment could be repeated here — for the edges, the unevenness of the screen, produce either violet and blue or yellow and yellow-red, and according to the way in which the screen is placed, they will spread across it to a greater or lesser extent. If that experiment proved nothing, nor does this one, and we shall spare our readers the *ergo bibamus* which normally follows at such a juncture.

Eleventh experiment

544.

This is where the author presents the main experiment, which we have referred to on various occasions, and which we have discussed in Chapter 19 under the heading of the combination of objective and subjective experiments (O. 350–355). It involves an image which is objectively projected on to a wall, then subjectively lowered, rendered colourless and coloured again in the reverse order.[49] Newton takes good care not to mention this experiment in its proper context for, since it makes nonsense of his theory, there is in fact no proper context for it in his book. It reduces, annihilates and reverses his ready-made and eternally immutable colours, and brings the modulation, continual development and constant mobility of the prismatic colours vividly before the senses. He, understandably therefore, only mentions the experiment in passing as a means of obtaining white light and of operating in it with combs. Although he describes it just

---

[49] See Figure 8 in the Appendix.

as we have described it, he maintains, as one might have expected him to, that the white of the image which has been subjectively lowered arises from the combination of all the coloured lights, since here, as in all prismatic experiments, complete whiteness marks the point of neutrality — where the boundary colours are on the point of switching over into their opposites. Within this image, which has turned white subjectively, he uses the teeth of his comb creating new stripes of colour by means of new obstacles; but externally, there being nothing internal about it.

Tenth experiment

545.

Here, we have a card with a really bent corner, an experiment which is, in fact, five or six experiments. Since we already know the value of all of them, however, since we are already convinced that none of them proves anything, we are not at all impressed by their being assembled and arranged in this way.

Instead of following the author word by word, as we have done hitherto, we therefore propose to give a short analysis of the various experiments from which this one has been compounded, the constituent parts of this monstrous whole, referring back to what has been said about them, individually and rounding off by commenting upon the experiment now in question.[50]

Analysis of the tenth experiment

546.

(1) A spectrum is formed in the usual way.
(2) It is projected on to a lens and intercepted by a white screen. A colourless round image forms at the focus.
(3) This image is subjectively displaced downwards, becoming coloured again.

---

[50] See Figure 6 in the Appendix.

(4) The screen is tilted. The colours appear just as they do in the second experiment of this part.

(5) Use is made of a comb: see the twelfth experiment of this part.

547.

Those interested will check for themselves the way in which Newton describes this complicated experiment and what he infers from it. Those who take the trouble to carry out the whole series of experiments again will be surprised and astonished by the total pointlessness of the way in which he amasses and arranges them. Nevertheless, since this experiment combines lenses and prisms, we shall return to it in our supplementary treatise.[51]

Thirteenth experiment

See Figure 3 Plate XIV

548.

In the foregoing experiment the several intervals of the Teeth of the Comb do the Office of so many Prisms, every interval producing the Phenomenon of one Prism.

549.

Certainly, but why? Because the multiple boundaries of the comb or rake produce fresh boundaries within the white region displayed in the refracted image of the large prism, and so give rise to the active entity of the regular play of colours.

550.

Whence instead of those Intervals using several Prisms. I try'd to compound Whiteness by mixing their Colours, and did it by using only three Prisms, as also by using only two.

---

[51] Newton performs a variant of this experiment, recreating white by substituting the lens with two prisms; see Figure 7 in the Appendix. It is ignored by Goethe; no doubt because he would have trouble trying to explain it according to his own theory.

551.

We simply note in passing, without dwelling at length on it, that an experiment involving several prisms is not the same as one involving a comb. As can be gathered from his figure and description, Newton uses no more than two prisms. Now let us see what is produced by means of them or, rather, between them.

552.

Let two Prisms ABC and abc, whose refracting Angles B and b are equal, be so placed parallel to one another, that the refracting Angle B of the one may touch the Angle at the Base Of the other, and their Planes CB and cb, at which the Rays emerge, may lie in Directum. Then let the Light trajected through them fall upon the Paper MN., distant about 8 or 12 Inches from the Prisms. And the Colours generated by the interior Limits B and c of the two Prisms, will be mingled at PT, and there compounded white.

553.

We shall deal with this paragraph, which contains much that is dubious, by analysing it from end to beginning. Newton admits here what he has so often obstinately denied, namely, that colours appear at boundaries, and as is typical of him, he does so in passing. It is not unreasonable to ask therefore, why on this occasion he operates so close to the prisms, why there are only 8 or 12 inches between them and the screen. He does not admit it, of course, but the reason is that at this distance the white he wants to produce is still present in its original form, the coloured borders at the edges still being so narrow that they are unable to overlap and create green. Newton is therefore not justified in drawing five lines spreading out from the edges B and c, as if two complete spectral systems were emerging there. All that can emerge at c is a blue and blue-red fringe, and at B nothing but a yellow-red and yellow one. Another important point, not to be overlooked, is that if one carries out the experiment in accordance with Newton's description and not his figure, so that the angles B and c are in immediate contact and the sides CB and cb lie in one line, no colours can emerge from the points B and c, since glass is touching glass, what is transparent combining with that which is transparent; and there are therefore no boundaries.

554.

Nevertheless, since we can agree with Newton when he goes on to maintain that the phenomenon occurs when the edges B and c are not in immediate contact, we have to give careful consideration to what then takes place, for this is a context in which his erroneous doctrine diverges from what is true. The initial appearance is in a state of nascency; as has already been observed, blue and blue-red develop at c, yellow-red and yellow at B. If these colours are now superimposed upon each other on the screen, the blue cancels out and neutralises the yellow-red, the blue-red, and the yellow. Since the rest of the area between M and N, in which the other coloured borders appear, is still white, it follows that, since the region where the coloured edges superimpose upon each other is white, the whole area must therefore appear white.

555.

If we move the screen further back, so that the spectrum is fully developed, green appearing in the middle, we shall no longer be able to form colourless areas by superimposition. Unlike the colours just mentioned, this spectrum is created by the displacement of a bright image, and can neither be neutralised by itself nor by the action of a second, similar image. If the second spectrum is lowered so that it is superimposed upon the first, the combination of yellow-red and blue-red produces purple; the combination of yellow-red and blue should create a colourless area, but since blue is related to green and the remainder is incorporated in the violet, decisive neutralisation is impossible. Nor is the green cancelled out when yellow-red is superimposed upon it for, at best, it is only its blue component which is countered, the yellow being augmented. It goes without saying that yellow-red is strengthened when it is projected on to yellow and yellow-red. And it is thus quite clear from all this how two full spectra interrelate when they are partly or fully superimposed.

556.

If one wishes to cancel out the middle of such a complete spectrum, that is to say, do away with the green, one will only be able to do so by forming complete spectra with two prisms, generating purple by uniting the yellow-red of one with the violet of the other, and projecting this on to the place occupied by the green of a third complete spectrum. This area will appear colourless, bright, and might be said to be white, for within it the true totality of colours unifies itself and cancels out anything specific. It is only natural that in such a place one might not notice the σχιερόυ for the colours being projected there have three colour images, a triple illumination, behind them.

557.

We should not miss this opportunity of mentioning the ingenious idea that one might attempt to remove the usual yellowness of lamplight and render it colourless by imparting a slightly violet tone to the glass cylinder of the Argand lamp.[52]

558.

Here, we have the truth of the matter. It cannot be denied that this is the appearance: compare our explanation and derivation of it with Newton's; ours will be found to be entirely adequate in every situation — his only under certain miserably contrived conditions.

Fourteenth experiment

559.

Hitherto I have produced Whiteness by mixing the Colours of Prisms.

560.

We have already dealt in detail with the extent to which he succeeded in producing this white.

---

[52] This was the first scientifically designed oil lamp, patented in England in 1784.

561.

If now the Colours of natural Bodies are to be mingled, let water a little thicken'd with Soap be agitated and raise a Froth, and after that Froth has stood a little, there will appear to one that shall view it intently various Colours everywhere in the Surfaces of the several Bubbles; but to one that shall go so far off, that he cannot distinguish the Colours from one another, the whole Froth will grow white with a perfect Whiteness.

562.

Anyone who is able to switch from refraction to epoptics — two totally different topics — in this way must be of a disposition and attitude of mind which is capable of accepting uncritically whatever turns up. Of the manifold objections to be raised against this experiment we shall mention only one: the distinguishable is contrasted here with the indistinguishable, but something does not cease to be, does not cease to exist in a third entity, because it can no longer be perceived by the external senses. If a dress has small marks on it, it does not become clean just because at a certain distance I can no longer see them — the paper does not turn white because I am too far away to distinguish the tiny lines drawn upon it. The chemist, by means of his reagents, renders visible certain parts of the most dilute infusions which healthy, straightforward observation could never have detected. And in Newton's ease we have evidence not of healthy and straightforward observation, but of a forced and prejudiced approach bent on bolstering certain presuppositions, as is evident in the next experiment.

Fifteenth experiment

563.

Lastly, in attempting to compound a white, by mixing the coloured Powders which Painters use, I consider'd that all Colour'd Powders do suppress and Stop in them a very considerable Part of the Light by which they are illuminated.

564.

Here, once again we have the author's preconception which, like his second thoughts, we are by now familiar with. He is obliged to acknowledge that colour involves darkness, but, since he does not know what to make of this, he starts to peddle his dubious experiments and erroneous conclusions again, so that the prospect becomes ever more dismal and depressing.

565.

For they become colour'd by reflecting the Light of their Colours more copiously; and that of all other Colours more sparingly, and yet they do not reflect the Light of their own Colours: so copiously as white Bodies do. If red Lead, for instance, and a white Paper, be placed in the red Light of the colour'd spectrum made in a dark Chamber by the Refraction of a Prism as is described in the third Experiment of the first Part of this Book; the Paper will appear more lucid than the red Lead, and therefore reflect the red-making Rays more copiously than the red Lead doth.

566.

The rashness of this last conclusion is typical of Newton's approach. The white is a bright base which works back through the red half-light by which it is illuminated red to be seen with complete clarity; the red lead, however, is a somewhat darker base, for although it is certainly similar to the prismatic red in colour, it does not have the same specifications. Although it also works back through the red prismatic half-light by which it is illuminated, it does so as a semi-dark base. As is only natural, therefore, the result is a colour which is strengthened, duplicated, and darkened.

567.

And if they be held in the Light of any other Colour, the Light reflected by the Paper will exceed the light reflected by the red Lead in a much greater Proportion.

568.

And this is in accord with Nature, as we have already made absolutely
clear. For all the colours appear on the white paper, each in accordance with its own determination, and without the mixing, interfer-
ence, and darkening brought about by the red lead when it is
displaced towards the yellow, green, blue, and violet. Since our readers are already well aware that the other colours behave in the same
way, rather than being surprised by the following passage, they will
almost certainly derive some amusement from the cross-grained but
self-assured manner in which he now continues.

569.

*And the like happens in Powders of some other Colours. And therefore by
mixing such Powders, we are not to expect a strong and full White, such
as is that of Paper, but some dusky obscure one, such as might arise from
a Mixture of Light and Darkness.*

570.

Here, at last, he lets slip the expression he has held back for so long;
like Balaam he has to bless whenever he wants to curse, and all his
stubbornness is of no avail against the demon of truth, which so often
stands in his way when he wants to set forth on his ass. From light
and darkness! We wanted nothing more. It is from light and darkness
that we have derived the formation of colours, and what pertains to
each singly, in its particular specificness, as its main characteristic,
what pertains to all the juxtaposed colours as the characteristic they
have in common, will also pertain to the mixture within which the
specific characteristics disappear. We therefore do not object at all to
the way in which he continues:

571.

*Or from white and black, that is, a grey, or dun, or russet: brown, such
as are the Colours of a Man's Nail, of a Mouse, of Ashes, of ordinary
Stones, of Mortar, of Dust and Dirt in High-ways, and the like. And
such a dark white I have often produced by mixing colour'd Powders.*

572

Which no one is likely to question; and what a blessing it would be were all Newtonians to be distinguished from the more enlightened by being obliged to wear body linen to match!

573.

There can be no doubt that he does manage to compound a black-and-white from coloured powders; now let us see how he attempts to form a grey which is at least as bright as it can be.

574.

For thus one Part of red Lead, and five Parts of Viride Aeris, composed a dun Colour like that of a Mouse.

575.

Newton begins by using pulverised verdigris, because of its bright mealy appearance, and indeed he always takes care to avoid using saturated colours.

576.

For these two Colours were severally so compounded of others, that in both together were a Mixture of all Colours.

577.

Here, he attempts to avoid the objection that he has not compounded his non-colour from all the others. We shall learn later from the *History* of the subsequent controversy among physicists concerning the mixing of colours in general and the ultimate composition of the non-colour — it being derived from three, five, or seven colours.

578.

Again, one Part of red Lead, and four Parts of blue Bise, composed a grey Colour verging a little to purple, and by adding to this a certain Mixture of Orpiment and Viride Aeris in a due Proportion, the Mixture lost its purple Tincture, and became perfectly grey.[53] *But the Experiment succeeded best without Minium thus. To Orpiment I added little by little a certain full bright purple, which painters use, until the Orpiment ceased to be yellow, and became of a pale red. Then I diluted that red by adding a little Viride Aeris, until it became of such a grey or pale white, as verged to no one of the Colours more than to another. For thus it became of a Colour equal in Whiteness to that of Ashes, or of Wood newly cut, or of a Man's Skin.*

579.

This mixture, too, has as its main ingredients bice and verdigris, both of which have a mealy, chalky appearance. In fact all Newton had to do in order to thin the colours out and produce lighter grey was to go on adding chalk, although nothing at all would have been proved by doing so.

580.

Now, considering that these grey and dark[54] *Colours may also be produced by mixing Whites and Blacks, and by consequence differ from perfect Whites, not in Species of Colours, but only in degree of Luminousness.*

581.

Here we have yet more of his insidious scheming, evidence of an attitude of mind which will have to be dealt with elsewhere, and which at present we shall only touch upon. Imagine a fully illuminated sheet of white paper in bright sunlight being placed in the shade. Consider what happens as daylight gradually turns into twilight until, finally, the paper is enveloped in darkness and is no longer visible. As the effectiveness of the light decreases, so does the counter-effect of the paper, and in this sense we can picture the white as making a gradual transition into black. Nevertheless, the course of the phenomenon can be said to be of a dynamic or ideal nature.

---

[53] Newton's word was "dun".
[54] Again, Newton's word was "dun".

582.

The situation is quite the reverse if we think of a sheet of white paper, in the light, which we first coat with a thin black tincture. By proceeding to give it second and third coats we deepen the darkness of the grey so much that the paper is as black as we can make it, nothing of the white base still shining through. Thus, seen atomistically or mechanically, a real darkness has been spread over the paper — which to some extent can be conditioned and moderated by the incident light; but it cannot be removed by it. Our sophist now attempts to present the matter as partaking of both situations, as an intermediate case, in the interpretation of which he is justified in taking both aspects into consideration. As we shall now see, this in fact means that he confuses them.

583.

It is manifest that there is nothing more requisite to make them perfectly white than to increase their Light sufficiently; and, on the contrary, if by increasing their Light they can be brought to perfect Whiteness, it will thence also follow, that they are of the same Species of Colour with the best Whites, and differ from them only in the Quantity of Light.

584.

It is a major shortcoming, not only in Newtonian optics but also in various other works, and particularly in those published during the period in question, that those who put pen to paper are unsure which standpoint to adopt, that they may at one moment have their feet firmly planted in reality, then adopt an idealistic stance, only to drift back once more into reality. This leads to the strangest kinds of presentation and explanation — one cannot deny that they have a certain content, but their form does involve an inner contradiction. This can be seen in the way in which Newton now attempts to elevate his light grey to white.

585.

And this I tried as follows. I took the third of the above mention'd grey Mixtures (that which was compounded of Orpiment, Purple, Bise, and

*Viride Aeris) and rubbed it thickly upon the Floor of my Chamber, where
the sun shone upon it through the opened Casement; and by it in the
shadow, I laid a Piece of white Paper of the same Bigness.*

586.

What has our worthy friend done now? In order to compare two
things and get them to cancel each other out, he has to remove the
difference existing between them. It is like placing a child on a table
beside a man, and then maintaining that they are the same size.

587.

The white paper in the shadow is no longer white: for it is darkened,
shaded; the grey powder in the sun is not white either: for its duski-
ness is inextinguishable. Since we are now aware of the ridiculousness
of the set-up, let us see what our observer makes of it.

588.

*Then going from them to the distance of 12 or 18 Feet, so that I could not
discern the Unevenness of the Surface of the Powder, nor the little
Shadows let fall from the gritty Particles thereof; the Powder appeared
intensely white, so as to transcend even the Paper itself in Whiteness,
especially if the Paper were a little shaded from the Light of the Clouds,
and then the Paper compared with the Powder appeared of such a grey
Colour as the Powder had done before.*

589.

What could be more natural? If the paper with which the powder is
to be compared is progressively darkened by an increasingly distinct
shadow, it is most certain that it will gradually become greyer. It is,
however, when he places the paper beside the powder in the sun, or
when he sprinkles his powder on white paper and places it in the sun,
that the true relationship becomes apparent.

590.

The rest of what he puts forward adds nothing to his argument, and we shall not enter into it. Eventually, he is actually visited by a friend, who also pronounces the grey powder laying in the sun to be white, as anyone might easily do who is suddenly asked to testify concerning things which have an ambiguous effect upon the senses.

591.

We shall disregard his triumphant *ergo bibamus*, since for those prepared to adopt the right approach, enough has already been said.

Proposition VI. Problem II

In a mixture of Primary Colours, the Quantity and Quality of each being given, to know, the Colour of the Compound

592.

All will agree that a colour scheme can be conveniently represented as a circle, and the first figure of our first plate illustrates what we considered to be the most advantageous way of doing so. Newton now proposes to do the same; but how does he set about it? The familiar flame-like spread of spectral colour is to be bent into a circle, while the areas the colours occupy on the periphery are determined according to musical intervals, which Newton claims to have discovered in the spectrum.

593.

But now a new difficulty arises; since all shades of red have to be specified between his violet and orange, he has to include within his circle of primary colours the pure red which is absent from his spectrum. Nevertheless, he only needs to make one of his small adjustments, to indulge in a little shady business, in order to interpolate this red in the same way that he did the green and white. Centres of gravity are now postulated together with little circles in certain proportions, and lines are drawn indicating the particular colour which results from the mixing of various primaries.

594.

We must leave our readers to study the author's account of this latest flight of fancy for themselves. We do not propose to concern ourselves with it, since we find it only too obvious that there is nothing natural about the spatial distribution of the colours around the proposed circle, the correlation between the spectrum and the musical scale being entirely fictional; it is, moreover, quite impossible to derive the opposing or mutually evocative colours from this Newtonian circle. Incidentally, after he has done with his measurings and lettering, he says himself that he conceives of this rule as: "accurate enough for practice, though not mathematically accurate". Since the practical usefulness of this colour scheme is nil, and since it is devoid of any valid theoretical foundation, what is the point of it?

Proposition VII. Theorem V

All the Colours in the Universe which are made by Light, and depend not on the Power of Imagination, are either the Colours of homogeneal Lights, or compounded of these, and that either accurately or very nearly, according to the Rule of the foregoing Problem

595.

Under this rubric Newton recapitulates the thesis he has defended throughout this second part of the first book. As the proposition indicates, his conclusion is that all colours pertaining to bodies are in fact simply integrating parts of light, forced, harassed, or separated out of it in various ways before being mixed up together again. Since we have examined the contents of this second part step-by-step, there is no need for us to dwell upon this repetition.

596.

He concludes by mentioning the colours we have classified and dealt with as being physiological and pathological. These are supposed not to pertain to light and, since he wants to dispense with them once and for all, he attributes them to the power of the imagination.

Proposition VIII. Problem III

By the discovered Properties of Light to explain the Colours made by Prisms

597.

Is it not curious in the extreme that this problem should arise at this juncture? From the very beginning of his *Opticks*, Newton has been concerned with investigating prismatic colours in order to discover the properties of light. Had he succeeded in doing so, nothing would have been easier than to reverse the procedure and derive the colours of the spectrum from the revealed properties of light.

598.

The root of the matter is that scheming is involved in this problem, too. In the figure in question, the twelfth of his second part and the ninth in our seventh plate, he gives the first clear indication of the unchanged white which occurs between the two coloured fringe phenomena. It has, however, been introduced unobtrusively on several previous occasions, the last being experiment 13, in which he used two prisms. Both there and here he represents the fringe phenomena as a series of five lines, by which he would appear to imply that the whole system of colours always emerges at both ends. More carefully considered, however, we can see that he has at last recognised the fringe phenomena with which we are so familiar; yet, instead of acknowledging the simple tendency of these phenomena to merge and form green, he makes the extraordinary claim that the colours deploy themselves as a series, whilst also overlapping and mixing with one another. From this medley of words and lines he then claims to have produced white, which is certainly present in the appearance, but which is there in its own right, not as the result of it having sprung from the coloured and overlapping lights of his hypothesis.

599.

The way he sees the matter is as confused and abstruse as it is erroneous: although he attempted to make it intelligible by employing Greek as well as Latin letters, he failed to manage it, and the faithful and devout among his followers have therefore found it necessary to transform his linear representation of it into a tabular one.

600.

Gren of Halle,[55] in the course of condemning our innocent contributions to optics with all the pride and fervour of a priest, worked out such a lettered tabular presentation of the displacement of a bright image. In the review of our *Contributions* published in the *Jena Literary Journal* the same was done for the displacement of a dark image. Since not everyone is able to give the required attention to this hotch-potch of letters, however, we have devoted plates 9 and 10 of this present work to provide the reader with a clear presentation of it. The prismatic colour systems are shown as coloured squares deployed in successive vertical sequences, partly together, partly in divided sections and detachments. They can, therefore, be compared horizontally at a single glance, so that by using straightforward common sense one is able to assess how ridiculous the results would be had Newton and his school grasped the truth of the matter.

---

[55] Friedrich Gren, a German physicist contemporary with Goethe.

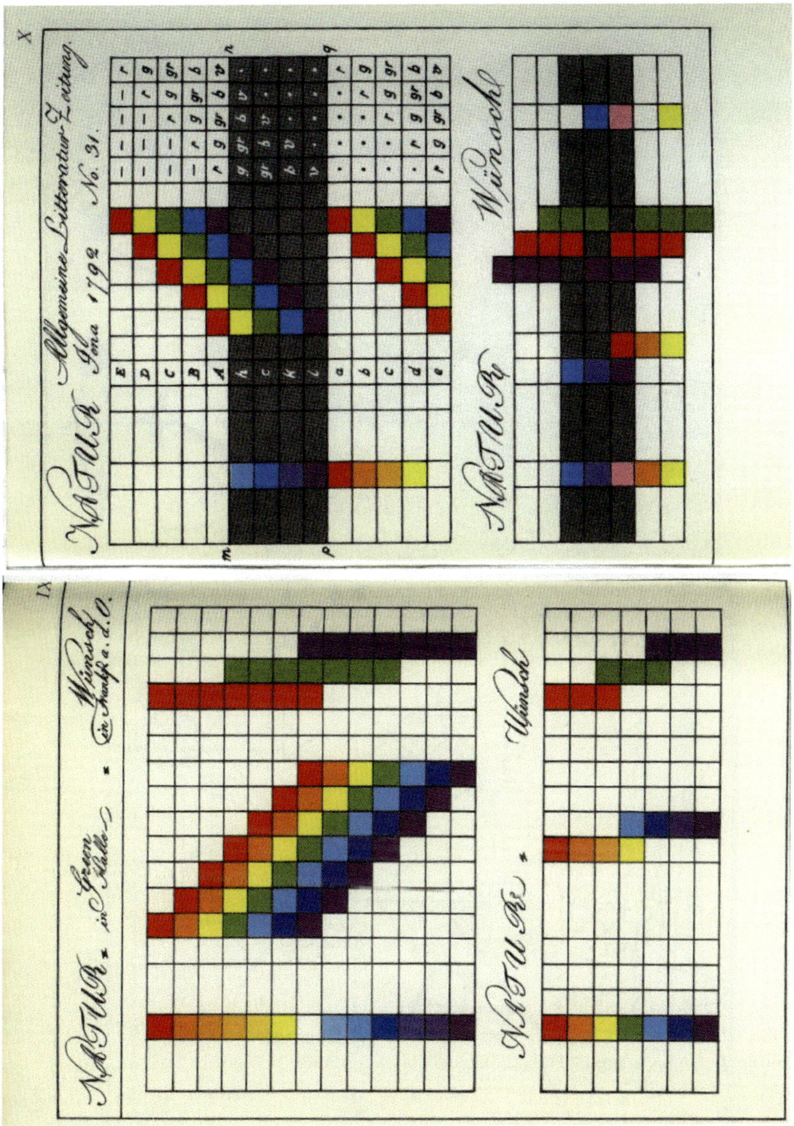

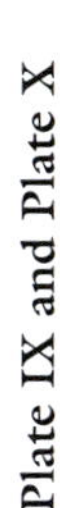

Plate IX and Plate X

601.

We have also included other colour series in these plates in order to illustrate the peculiar reduction of the prismatic colour phenomenon by Wünch. This strange person attempted to vindicate Newton's presentation by producing his own epitome of it; he indulged in the strangest ideas and believed he had simplified the matter, although he had merely succeeded in making it more unnatural.

602.

We shall return to this theme when we provide our explanation of the plates, when our readers will surely allow us the liberty of taking a reasonable reward for our labours, and having a good laugh at these opponents and semi-opponents, whilst not entirely forgetting their master.

Sixteenth experiment

603.

This phenomenon is taken up in a merely empirical manner: rather than being properly illustrated, it is, as it were, simply appended to the preceding hypothetical procedure by means of a crude figure, the thirteenth of the second part. If its true significance is to be grasped, it has to be made the subject of a proper experiment.[56]

604.

A prism is held in front of an open window with its refracting angle pointing downwards in the normal way, the observer leans out sufficiently far to avoid the upper framework of the window showing in the prism through refraction; under a dark edge at the top of the prism he will then notice a yellow arc spreading from the bright sky.

---

[56] See Figure 13(Book one Part II) in the Appendix.

This dark edge derives from the outer, upper edge of the prism, as can be easily confirmed by covering it with a piece of wax, which is then quite distinctly visible within the coloured arc.

The clear sky can then be seen below this coloured arc, and lower down the houses or mountains or whatever constitutes the horizon, all of which, in accordance with the law governing the matter, appears to be fringed in blue and blue-red.

The prism is now tilted further down, whilst one continues to look into it. The buildings constituting the horizon gradually fall away until they finally disappear, whilst the yellow and yellow-red arc visible hitherto switches into a blue and blue-red one — the one Newton investigates without mentioning the lead-up to the change in colour we have just described.

605.

Since this, too, is simply an empirical phenomenon and not an experiment, we shall now make an observation and then go straight on to suggest an arrangement for removing what is incidental to this appearance whilst also multiplying and confirming its conditions. The phenomenon, as it appears to us at the window, is the result of the bright sky being above the dark earth. We cannot easily reverse this in order to have a dark sky and a bright earth. The same is true of rooms, where the ceilings are usually bright and the walls more or less dark.

606.

With this in mind, the following arrangement is setup in a moderately large, high ceilinged room. At the top of the wall a strip of black paper is fixed next to a strip of white paper, whilst the corresponding strips fixed to the ceiling are white and black; in the angle between the wall and the ceiling one therefore has a clash of black and white and of white and black. These strips are then observed through the prism, side-by-side and one above the other, in the same way as was done when looking out of the window in the previous exercise. The arc reappears, although it now has to be distinguished from all the others which give rise to edges or stripes. The arc appears yellow where it covers the white strip on the ceiling, just as it did formerly when it covered the white sky. Where it covers the black strip, however, it appears blue. If the prism is tilted down again, so that the wall seems to fall away, the arc suddenly switches as it passes over the converse strips on the wall. Here, too, it appears yellow where it covers the white strip and blue where it covers the black one.

607.

Once one is aware of this, one will be able to recognise the phenomenon by means of chance empiricism — when walking through a snow-covered area, for instance, or following light sandy paths edged by dark grass verges. Since an extensive essay and a separate plate would be required for full clarification of the appearance, we shall limit ourselves to providing a full explanation of it. It is by this refraction, which draws downwards the objects immediately in front of us, that the objects or surfaces above us are driven toward the upper edge of the prism possibly as the result of an interfering reflection, and depending on how bright or dark they are, they are in accordance with the known laws. The edge of the prism appears as an arc, a form assumed by all horizontal lines lying in front of us when they are viewed through a prism.

Proposition IX. Problem IV

By the discovered Properties of Light to explain the Colours of the Rainbow

608.

It is only natural that all that applies to prisms should also apply to lenses; it is also to be expected that all that applies to spherical sections will also apply to spheres, despite the presence of a few further measurements and conditions. Consequently, Newton is acting in complete conformity with his theoretical and hypothetical procedure when he applies his theory concerning prisms and lenses to spheres and drops.

609.

Having taken issue with him over everything so far, it is only natural that we should contradict him here, too, and explain the phenomenon of the rainbow in our own way. It is a matter of applied physics, however, and we shall not concern ourselves with it here. We shall present whatever we find necessary to say about it in one of the supplementary articles.[57]

---

[57] Goethe does not do this — although he does carry out a correspondence with Sulpiz Boisserée about the rainbow in 1832. See *J. W. Goethe, Farbenlehre*, edited by G. Ott and H. Proskauer, 3 vols (Stuttgart, 1979), III, 273.

Proposition X. Problem V

By the discovered Properties of Light to explain the permanent Colours of Natural Bodies

610.

These Colours arise from hence, that some natural Bodies reflect some sorts of Rays, whereas others reflect other sorts more copiously than the rest.

611.

Take particular note of *more copiously*; no mention is made of their merely reflecting some rays or of their only reflecting others, as would have to be the case, at least with certain pure colours. On observing a pure yellow one might well acquiesce in the notion that it is emitting only the yellow rays; likewise with pure blue. But the author is careful to avoid committing himself to this since, as we shall now see, he has to keep the back door open again if he is to evade a serious challenger.

612.

Minium reflects the least refrangible or red-making Rays most copiously, and thence appears red. Violets reflect the most refrangible most copiously, and thence have their Colour, and so of other Bodies. Every Body reflects the Rays of its own Colour more copiously than the rest, and from their excess and predominance in the reflected Light has its Colour.

613.

The Newtonian theory has the characteristic of being very easy to learn and very difficult to apply. Grant the initial proposition with which the *Opticks* begins, believe in it, and one's mind will be set permanently at rest concerning the essence of colour. Undertake closer investigation, however, attempt to apply the hypothesis to the phenomena, and the difficulties begin. There are objections *a priori* and *a posteriori*. The acute observer comes across so many limitations, restrictions, and reservations, that he has to regard each of the propositions and then finally the doctrine as a whole as totally neutralised. What now follows is further evidence of this.

Seventeenth experiment

614.

For in the homogeneal Lights obtained by the solution of the problem proposed in the fourth Proposition of the first Part of this Book.

615.

It has been shown at the relevant juncture that, despite all the trouble taken, the lights obtained there were no more homogeneous than those obtained in the ordinary prismatic experiment.

616.

If you place Bodies of several Colours, you will find, as I have done that every Body looks most splendid and luminous in the Light of its own Colour.

617.

There can be no objection to this. The same effect is produced when the experiment is performed with the perfectly normal and undistorted prismatic image, however, and nothing is more natural than that an effect should be enhanced rather than toned down when like is added to like, when the one homogeneity is a degree more effective than the other. If common vinegar has concentrated vinegar added to it, the mixture will be stronger than the original. It is quite different when something heterogeneous is added, when common vinegar is mixed with alkali. The effect of both is then lost, they neutralise one another. But Newton will and can know nothing of this homogeneity and heterogeneity. Although he agonises over his degrees and gradations, he eventually has to acknowledge an opposite effect.

618.

Cinnaber in the homogeneal red light is most resplendent, in the green Light it is manifestly less resplendent, and in the blue Light still less.

619.

What a poor description of the phenomenon this is; he considers nothing but the cinnabar and its resplendence, does not mention the mixture which arises from the incident prismatic colour and the corporeal colour of the underlying body.

620.

Indigo in the violet blue light is most resplendent.

621.

And why should this be so? The reason is that indigo, which is in fact only a dark saturated blue colour, is endowed by the violet light with a resplendence, a lustre, with brightness and vivacity and, as it is exposed to green, yellow, and red, its resplendence is gradually diminished.

622.

Now why does the author only mention the resplendence, which is supposed to diminish? Why does he not mention the new appearance of mixed colours which is formed in this way? Since the truth is too natural, use is made of what is false or half true in order to lend lustre to the unnaturalness of what the matter has been turned into.

623.

By a leek.

624.

And, what is the point of experimenting with garlic after one has been using a powder? Why does he not keep to even surfaces such as paper or a taut silk material? It could be that the garlic is meant to indicate that the doctrine is also applicable to plants.

625.

The green light, and next that, the blue and yellow, which compound green, are more strongly reflected than the other Colours red and violet.

626.

But to make these Experiments the more manifest, such Bodies ought to be chosen as have the fullest and most vivid Colours, and two of those Bodies are to be compared together. Thus, for instance, if Cinnaber and ultra-marine blue.

627.

We have often observed that one ought not to use powders; for how can their uneven parts be prevented from casting shadows?

628.

Be held together in the red homogeneal Light, they will both appear red.

629.

He only says this in order to retract it almost immediately.

630.

But the Cinnaber will appear of a strongly luminous and resplendent red, and the ultra-marine blue of a faint obscure and dark red.

631.

Perfectly understandable; for yellow-red enhances yellow-red and vitiates blue.

632.

And if they be held together in the blue homogeneal Light, they will both appear blue, but the ultra-marine will appear of a strongly luminous and resplendent blue, and the Cinnaber of a faint and dark blue.

633.

Also perfectly understandable in the light of our interpretation.

We are most reluctant to repeat these things here, since we have already dealt with them in detail above. One is obliged to repeat them, however, since in order to drive his points home more effectively, Newton constantly reiterates what is erroneous.

634.

Which puts it out of dispute that the Cinnaber reflects the red Light much more copiously than the ultra-marine doth, and the ultra-marine reflects the blue Light much more copiously than the Cinnaber doth.

635.

This is typical of the way in which he puts something beyond dispute — by making an unqualified statement of an opinion, an assertion concerning observations, the only basis of which is words and the way in which they have been employed. For in Newtonianism, the essence of colour is nothing more than verbiage; which is most effective in obscuring Nature.

636.

The same Experiment may be tried successfully with red Lead and Indigo, or with any other two colour'd Bodies (in order to perceive)[58] the different strength or weakness of their Colour and Light.

---

[58] Newton's own words were; "if due allowance be made for".

637.

Those who are well-informed are already aware of what is perceived here.

638.

And as the reason of the Colours of natural Bodies is evident by these Experiments.

639.

All that is evident is that he is adapting the appearance to his hypothesis by providing an incomplete and clumsy description of it.

640.

So it is farther confirmed and put past dispute by the two first Experiments of the first Part, whereby 'twas proved in such Bodies that the reflected Lights which differ in Colours do differ also in degrees of Refrangibility.

641.

The end is now linked artificially to the beginning: since we were cajoled there into believing that the corporeal colours are lights, it is understandable that these lights should now turn out to be fully fixed colours, further evidence for the validity of the argument.

No support for the theory can now be drawn from the experiments upon which this argument is based, however, since we have already provided a comprehensive refutation of the conclusions drawn from them.

642.

For thence it's certain, that some Bodies reflect the more refrangible, others the less refrangible Rays more copiously.

643.

And we are certain, not only that more refrangible rays and less refrangible rays do not exist, but that there is a more valid and convenient way of giving expression to natural phenomena.

644.

And that this is not only a true reason of these Colours, but even the only reason, may appear farther from this Consideration, that the Colour of homogeneal Light cannot be changed by the Reflection of natural Bodies.

645.

Newton must be very certain of the blind faith of his readers to assert that the colours of homogeneous light cannot be changed by reflection from natural bodies, for he has admitted on the previous page that red light reflects differently from cinnabar than from ultra-marine, that blue light reflects differently from ultra-marine than from cinnabar. It is now clear why his delivery is so contrived, why he only talks of the resplendence and brightness of the dullness and darkness of colour, but not of it being conditioned by mixing. He could hardly have spoken of such a clear and simple phenomenon in a more perverse and dishonest manner, in order to justify himself he certainly had to admit, consciously or unconsciously, like Reynard the fox that: "Lies are badly needed, and plenty of them, too!"

For, after having stated expressly that the prismatic colours undergo change on certain bodies, he continues as follows:

646.

For if Bodies by Reflection cannot in the least change the colour of any one sort of rays, they cannot appear colour'd by any other means than by reflecting those which either are of their own Colour, or which by mixture must produce it.

647.

Mixture now puts in an appearance, and in such a way that it is by no means clear what the purpose is; his conscience is evidently pricking him, and this is simply a transition to what follows, to his taking back all he has asserted. The reader should note well the author's incredible effrontery.

648.

But in trying Experiments of this kind care must be had that the Light be sufficiently homogeneal.

649.

In the straightforward prismatic experiment we have already thoroughly analysed his attempts to make the coloured prismatic lights appear more homogeneous than they are, so there is no need for us to repeat the procedure. The reader will recall that, in order to obtain something resembling the desired homogeneity, Newton had to prescribe the most difficult, not to say impossible, conditions. Note how he now casts suspicion upon a simple experiment, which can be performed by anyone, by continuing as follows:

650.

For if bodies be illuminated by the ordinary prismatic Colours, they will appear neither of their own daylight Colours, nor of the Colour of the Light cast on them, but of some middle Colour between both, as I have found by experience.

651.

It certainly is remarkable that, after finally admitting a certain experiment to be the only possible one, he should immediately proceed to cast doubt upon it. For what is true of the simplest prismatic appearance when it is projected on to corporeal colours remains true, regardless of its being forced and conditioned through holes of various sizes and by lenses of various focal lengths; nothing else can or will ever emerge.

652.

How does our author set about developing this uncertainty in the minds of his pupils? With extreme subtlety. And if one considers his action with an open mind, if one has a lively sense of what is true, one can certainly say that what he does is quite disgraceful. We are simply told that:

653.

Red Lead (for instance) illuminated with the ordinary prismatic green will not appear either red or green, but orange or yellow, or between yellow and green, accordingly as the green Light by which 'tis illuminated is more or less compounded.

654.

Why does he not proceed here by degrees or stages? If he projects perfectly ordinary prismatic red on to red lead, the latter will have just as fine and resplendent an appearance as it will if he uses the most forced of spectra. If he projects the green of such a spectrum on to red lead the appearance will be just as he describes it, or rather as we described it above when discussing the matter. Why, then, does he begin by casting doubt upon reasonable experiments? Why is he so excessively scrupulous? Why does he always end up by returning once again to the initial experiments? Simply in order to confuse people and leave an escape route for himself and his flock.

Reluctantly, we now draw attention to the typically distorted manner in which he attempts to explain the dissipation of prismatic green when it is projected on to red lead.

655.

For because red Lead appears red when illuminated with white Light, wherein all sorts of Rays are equally mix'd and in the green Light all sorts of Rays are not mix'd, (there must be a difference).

656.

Note that all sorts of rays are now supposed to be contained in green, and that this is at odds with his earlier exposition of the heterogeneity of homogeneous rays. There, he displays them within the imaginary circle and it is only the adjacent colours which intermesh, whereas here, since every colour pervades the whole image, there would appear to be no possibility of separating them out. Mention will eventually be made of further nonsense arising from this manner of representation, in which there can be a system of five, six, or seven systems ranged *in echelon.*

657.

The Excess of the yellow-making, green-making and blue-making Rays in the incident green Light, will cause these Rays to abound so much in the reflected Light, as to draw the Colour from red towards their Colour. And because the red Lead reflects the red-making Rays most copiously in proportion to their number, and next after them the orange-making and yellow-making Rays; these rays in the reflected Light will be more in proportion to the Light than they were in the incident light, and thereby will draw the reflected light from the green towards their Colour. And therefore the red lead will appear neither red nor green, but of a Colour between them both.

658.

Since we have already given a detailed explanation of the matter, it only remains for us to recommend the utter nonsense of this passage to posterity as a prime example of such a methodological approach.

He goes on to add four further observations, which he explains in his own way, and which we now intend to take note of and comment upon.

659.

In transparently colour'd Liquors 'tis observable, that their Colour uses to vary with their thickness, thus, for instance, a red Liquor in a conical

Glass held between the Light and the Eye, looks of a pale and dilute yellow at the bottom where 'tis thin, and a little higher where 'tis thicker grows orange, and where 'tis still thicker becomes red, and where 'tis thickest the red is deepest and darkest.

660.

We have illustrated this observation by using tapering jars (0. 517, 518), and have developed from it the important doctrine of augmentation: yellow, for example, like blue, tends toward red when it is denser and more darkened, and thus preserves the property we noticed as characteristic of its initial appearance in turbid media. Once we have grasped the profound simplicity of these archetypal and basic phenomena we are bound to find the way in which Newton interprets them particularly disturbing.

661.

For it is to be conceiv'd that such a Liquor stops the indigo-making and violet-making Rays most easily, the blue-making Rays more difficultly, the green-making Rays still more difficultly, and the red-making most difficultly: and that if the thickness of the Liquor be only so much as suffices to stop a competent number of the violet-making and indigo-making Rays, without diminishing much the number of the rest, the rest must (by Prop. 6. Part 2.) compound a pale yellow. But if the Liquor be so much thicker as to stop also a great number of the blue-making Rays, and some of the green-making, the rest must compound an orange; and where it is so thick as to stop also a great number of the green-making and a considerable number of the yellow-making the rest must begin to compound a red, and this red must grow deeper and darker as the yellow-making and orange-making Rays are more and more stopp'd by increasing the thickness of the Liquor, so that few Rays besides the red-making can get through.

662.

Can the history of the sciences provide us with any kind of explanation as ridiculous as this one?

663.

Of this kind is an Experiment lately related to me by Mr Halley, who, in diving deep into the Sea in a diving Vessel, found on a clear sunshine day, that when he was sunk many Fathoms deep into the Water the upper part of his Hand on which the sun shone directly through the Water and appeared of a red Colour, like that of a damask rose, and the Water below and the under part of his Hand illuminated by Light reflected from the water below look'd green.

664.

This experiment involves physiological colours, and it is in this context that we have already mentioned it. The water acts here as a turbid medium which gradually moderates the rays of the sun until they pass over from yellow into red and finally appear to be purple coloured; the shadows, however, are seen in the green colour evoked by this red. Note the curious way in which Newton now attempts to adapt his terminology to the phenomenon.

665.

For thence it may be gather'd, that the Seawater reflects back the violet and blue-making Rays most easily, and lets the red-making Rays pass most freely and copiously to great Depths. For thereby the sun's direct Light at all great Depths, by reason of the predominating red-making Rays, must appear red; and the greater the Depth is, the fuller and intenser must that red be. And at such Depths as the violet-making Rays scarce penetrate unto, the blue-making, green-making, and yellow-making Rays being reflected from below more copiously than the red-making ones, must compound a green.

666.

We are sufficiently aware of the true derivation of this phenomenon to be amused by Newton's conception of it; in that we also see the erroneousness of his explanation, however, his whole system will appear to be untenable.

667.

Now if there be two Liquors of full Colours, suppose a red and blue, and both of them so thick as suffices to make their Colours sufficiently full; though either Liquor be sufficiently transparent apart, yet will you not be able to see through both together. For if only the red-making Rays pass through one Liquor, and only the blue-making through the other, no Rays can pass through both. This Mr Hook tried casually with Glass Wedges filled with red and blue Liquors, and was surprised at the unexpected Event, the reason of it being then unknown; which makes me trust the more to his Experiment, though I have not tried it myself. But he that would repeat it, must take care the Liquors be of very good and full Colours.

668.

Now what is the basis of this whole experiment? All he is saying is that a medium which is only just transparent becomes opaque when doubled in thickness, and that this takes place regardless of there being one colour or two different colours, which are looked through separately and then in combination.

669.

This experiment has now haunted the history of the theory of colours for more than a century. In order to get to the bottom of it one needs certain glass containers, which we shall give a full description of when we discuss our apparatus. Several upright wedge-shaped jars are constructed from glass plates in such a way that they form parallelepipeds when placed together. One begins by filling them with pure water and familiarising oneself with the displacement of contrasting images and the well-known prismatic phenomena. One then slides two of them together and allows ink to drip in gradually until the liquid is opaque; finally, when one slides the two wedges apart, one will see that each by itself is still relatively transparent.

670.

If the same procedure is carried out with coloured liquors, the result is invariably the same irrespective of there being either one or two colours in the two vessels. As long as the liquids are not super-saturated one will be able to see quite well through the parallelepiped.

671.

One understands now why it is that Newton should lay such emphasis upon saturated and rich colours, for he does so both at the beginning and end of his treatise. In order to confirm that colour is in fact entirely irrelevant here, fill two of these wedge-shaped jars with a blue litmus so that the parallelepiped is still transparent. Get an assistant to drip vinegar into one of them so that its blue colour changes to red; the transparency will remain substantially the same and, since the σχιερόυ of the blue is removed by the acid, will even tend to increase somewhat. By diversifying the procedure, one can repeat all the experiments relating to the apparent mixing of colours.

672.

In order to impress the importance of these experiments upon oneself and others one needs to have four to six such jars to hand, in order not to waste time pouring from one to the other and to avoid incon-venience and mess. One should have no regrets about acquiring this apparatus for, with a little practice, good use can be made of it in performing objective and subjective experiments modified by col-oured media. We shall therefore repeat what we have already said above: a transparent medium becomes opaque when doubled or mul-tiplied in thickness, as is evident from coloured window panes, opal glasses, and even colourless window panes.

673.

Now Newton experiments with turbid media again. Since these archetypal phenomena have been dealt with in detail in the *Outline*,

we shall have no great difficulty in pointing out the inadequacy of the way in which he accounts for them.

674.

There are some Liquors, as the Tincture of Lignum Nephriticum, and some sorts of Glass which transmit one sort of Light most copiously, and reflect another sort, and thereby look of several Colours, according to the Position of the Light. But, if these Liquors or Glasses were so thick and massy that no Light could get through them I question not but they would like all other opake Bodies appear of one and the same Colour in all Positions of the Eye, though this I cannot yet affirm by Experience.

675.

And yet this particular experiment is easily performed. If a turbid medium which is still semi-transparent is held in front of a dark background it will appear to be blue. This blue is certainly not reflected from the surface, however, but comes out of the depths. If such bodies reflected blue from their surface more readily than any other colour, this surface would always appear to be blue, even if the degree of turbidity was increased until the medium became opaque. Yet, as is to be expected from causes given in the *Outline*, what one sees is white. Newton treats this as a particularly difficult matter, probably because he is uncertain of his ground.

676.

For all colour'd Bodies, so far as my Observation reaches, may be seen through if made sufficiently thin, and therefore are in some measure transparent, and differ only in degree of Transparency from tinged transparent Liquors; these Liquors, as well as those Bodies, by a sufficient Thickness becoming opake. A transparent Body which looks Of any Colour by transmitted Light, may also look of the same Colour by reflected Light, the Light of that Colour being reflected by the farther Surface of the Body, or by the Air beyond it. And then the reflected Colour will be diminished, and perhaps cease, by making the Body very thick, and pitching it on the backside to diminish the Reflection of its farther Surface, so that the light reflected from, the tinging Particles may, predominate, the Colour of the reflected Light will be apt to vary from that of the Light transmitted.

677.

If one is acquainted with the derivation of corporeal colours as we have attempted to present it in the *Outline* one will see no point in talking around the subject like this, especially if one is also convinced that a colour can only be visible if there is a light in the background, and that in actual fact no corporeal colour can be perceived by us in the absence of an incident light which either traverses a transparent body or penetrates to the bright base of an opaque body, from whence it is reflected. We shall disregard the author's *ergo bibamus* and hasten with him to the conclusion.

Proposition XI. Problem VI

***By mixing colour'd Lights to compound
a beam of Light of the same Colour and
Nature with a beam of the sun's direct Light,
and therein to experience the truth of the
foregoing Propositions***

678.

In this experiment Newton combines prisms and lenses again and we therefore refer our readers to the supplementary chapter — where the problem belongs. The long and short of it is that here, too, nothing is achieved; for he simply raises a colour phenomenon to its highest pitch with a prism, and then nullifies it with a lens; after passing through the lens it is inverted, blue and violet appearing at the bottom, yellow and yellow-red at the top. Thus fringed, the image is then projected on to another prism, thereby displacing the inverted image upwards, inverting it yet again and reducing the fringes to the point of nullity; this colourless image is then intercepted by a third prism, with its refracting angle pointing upwards, and after refraction appears coloured again.[59]

---

[59] See Figure 16 in the Appendix.

679.

We see nothing noteworthy in all this; it is no surprise to us that there are various ways in which one can start again and recolour such a decoloured image by means of a new displacement, without this new colouration having the slightest connection with that which has been removed; we have devoted our eighth plate, which concerns various kinds of reflection, to a detailed treatment of the subject.

680.

Attentive readers and experimenters will therefore be well aware that when such coloured images have been brought back to the point of nullity, either subjectively or objectively, they retain, maintain, renew, or reverse their characteristics, either in accordance with the laws governing the initial action, or in accordance with the opposite determination.

Conclusion

We believe we have now fulfilled our task of providing a polemical treatment of the first book of the *Opticks*, that we have clearly demonstrated the extent to which Newton's hypothetical explanation and derivation of the coloured phenomena caused by refraction are untenable. The succeeding books deal with what we have called epoptic and paroptic phenomena, and we have no intention of examining them. Historical accounts of Newton's explanation and exposition of these phenomena are to be found in all subsequent histories and dictionaries of physics. They have never been very influential, however, and since the contemporary scientific community, even including the author's own countrymen, no longer has any interest in them, there is no reason why we should explore the matter any further. Anyone who wants to do so should compare our exposition of epoptic phenomena with Newton's. Whereas we have traced them back to simple elements, he proceeds in his usual way, confusing what is important with what is incidental, and then proceeding to measure and calculate, explain and theorise on the basis of one or the other or all together, just as he does when dealing with refraction. Were we to criticise the succeeding books we would, therefore, have to repeat the procedures we have adopted in dealing with the first one.

As we look back over this work, we wish we were in the position of the cardinal who always had his writings printed in a preliminary draft. There is much that we would have revised and explained more satisfactorily. We would, perhaps, have toned down some of our sharper remarks, attempted to avoid provoking an opponent, annoying the uncommitted, obliging a friend to play the apologist.

Nevertheless, we draw some comfort from reminding ourselves that the work in its entirety was undertaken and completed in the midst of the mightiest war that has ever convulsed the fatherland. Unfortunately, the violence of the times invades even the peaceful dwellings of the muses and, although human behaviour may not be wholly determined by what is going on around, it is not modified by it. Over the years, experience and observation have taught us that when opinions or actions are in conflict, one's opponent has to be subdued not protected, that no one can be flattered or complimented into giving up an advantage, and that it may well be necessary to relieve him of it by force. There is no more stubborn party than the Newtonian in the whole history of the sciences. Since it has embittered the lives of many who have sought truth, and since for many years it has prevented me from undertaking more satisfying and useful work, I believe there may be some excuse for my having taken every opportunity to denigrate it and its author. I can only hope that this polemic will be of value to our successors.

This is not the end of the matter, however, for to a certain extent it will be taken up again in the historical part, where we shall have to show how such an exceptional person came to make such a mistake, how he was able to persist with it and persuade so many outstanding people to grant him their approval. This will involve more than merely polemicising, for as well as impeaching the perpetrator and his pupils, the century which approved of and persisted with the doctrine, we shall also have to absolve them. It is to this milder task, necessary as it is to the completion and conclusion of the whole, that we now invite our readers, in the hope that they will undertake it with an open mind and in good faith.

In Lieu of an Epilogue

Newtonian Doctrine

(1) Light is compound: heterogeneous.
(2) Light is composed of coloured lights.
(3) Light is decomposed by refraction, reflection, and inflection.
(4) It is decomposed into seven, or rather, innumerable parts.
(5) As it is decomposed, so can it be re-compounded.
(6) The observable colours do not arise from a modification of light from without or by external conditions.

Results of my experiments

(1) Light is the simplest, most elementary, most homogeneous entity that we know. It is not composite.
(2) Least of all is it compounded of coloured lights. Any light which has assumed a colour is darker than colourless light. Brightness cannot be compounded out of darkness.
(3) Reflection, refraction, and inflection are three conditions whereby we often experience observable colours; but all three are the occasion for their appearance rather than their cause. For all three conditions can occur without colour being present. There are moreover other conditions which are more important, as for instance the moderation of light, the interaction of light and shadow.
(4) There are only two pure colours, blue and yellow. There is a colour condition pertaining to them both, red; and two mixtures,

green and purple; the remainder are graduations of these colours, or impurities.

(5) Colourless light cannot be compounded from observable colours, nor out of coloured pigments. All such performed experiments are either flawed, or incorrectly carried out.

(6) The observable colours arise by modification of light by external conditions. The colours result from the excitement of light, they are not released from it. Once the conditions lapse, then light reverts to being colourless, as before, not because the colours retreat back into the light, but because they cease. Shadows become colourless when the effect of the second light is removed.

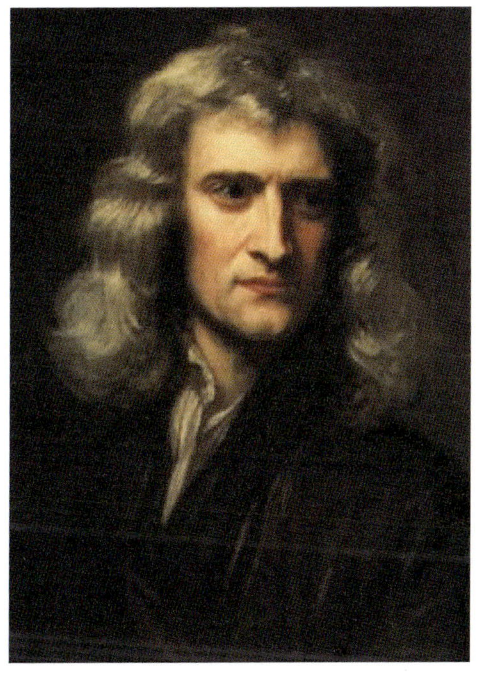

Sir Isaac Newton (aged 46)

A 1689 portrait by Godfrey Kneller

Reproduced by kind permission of the Trustees of the Portsmouth Estates

Appendix

Some of the figures relating to Newton's experiments, although treated by Goethe, were not reproduced by him in his Plates. So, for the benefit of those readers who do not have a copy of *Opticks* to hand, they are reproduced here. They are taken from the 4th edition (London, 1730). Reprinted by the Dover Press (New York, 1952).

Book 1, Part I

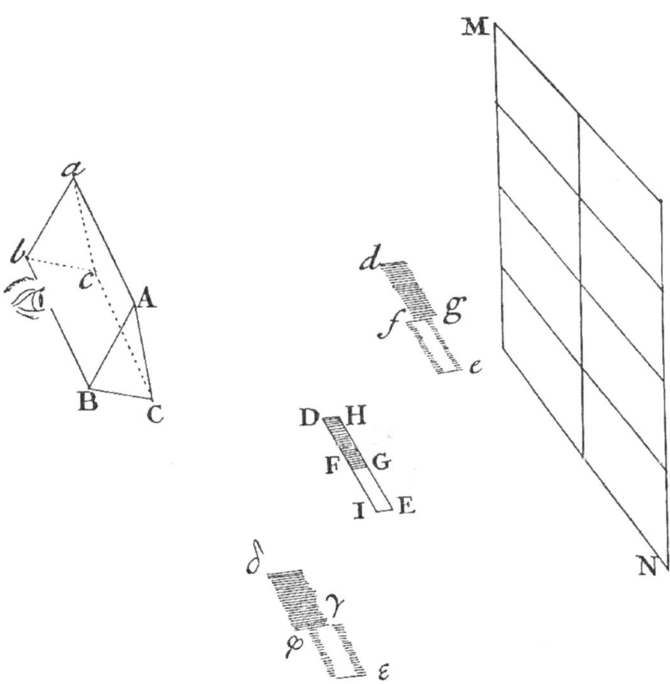

Figure 11   Referred to in paragraph 34.

                    Appendix

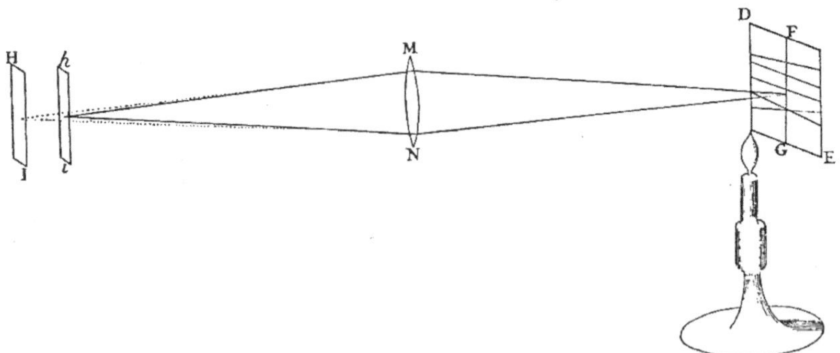

Figure 12   Referred to in paragraph 48.

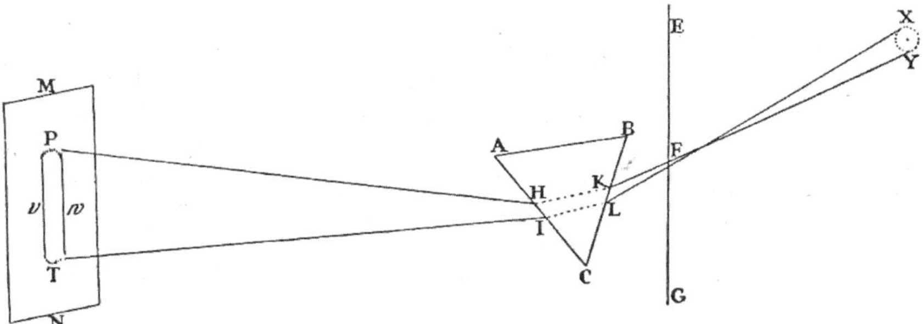

Figure 13   Referred to in paragraph 86.

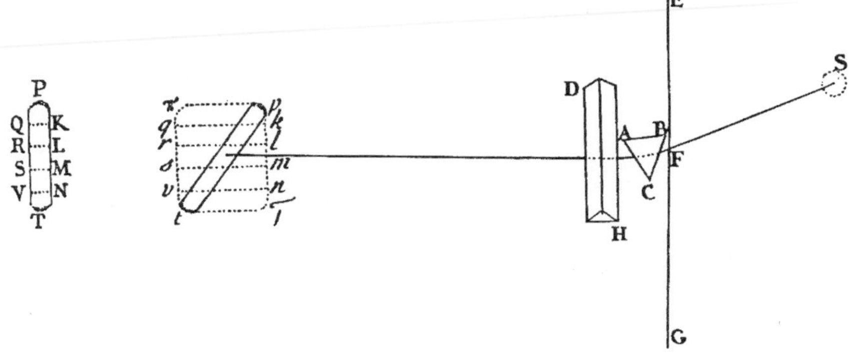

Figure 14   Referred to in paragraph 99.

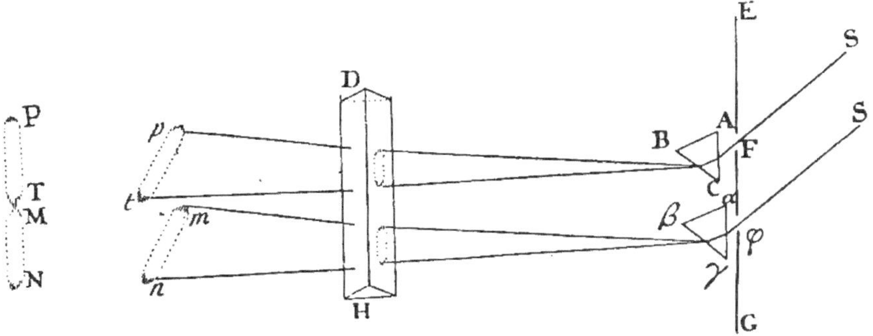

Figure 17   Referred to in paragraph 111.

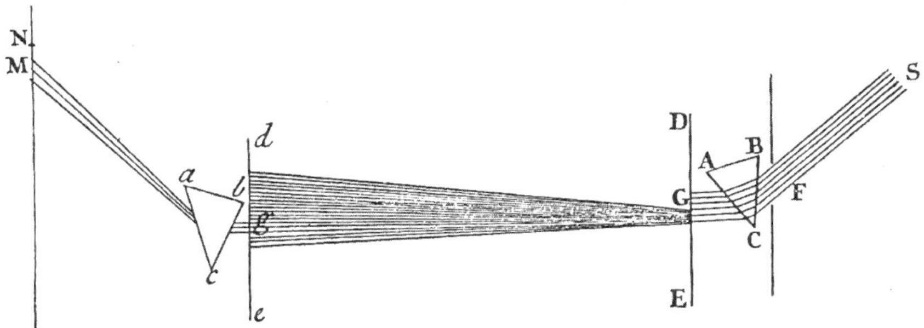

Figure 18   Referred to in paragraph 115.

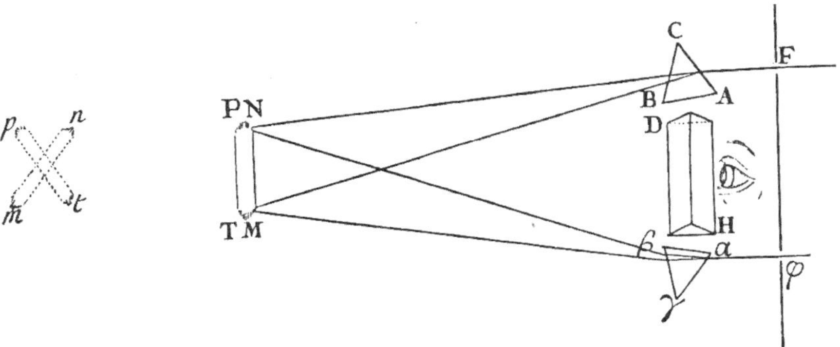

Figure 20   Referred to in paragraph 143.

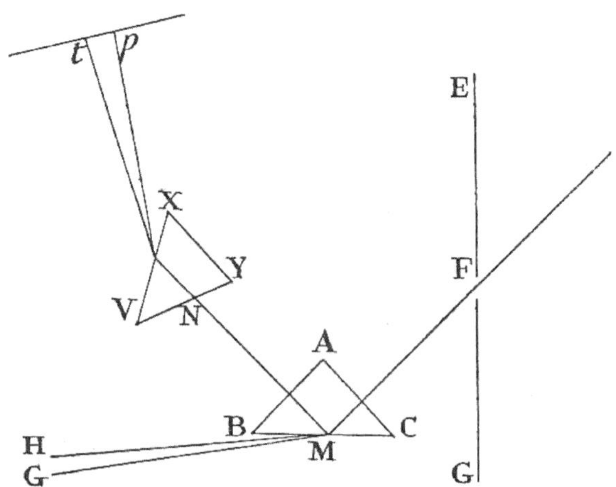

Figure 21   Referred to in paragraph 196.

Book 1, Part II

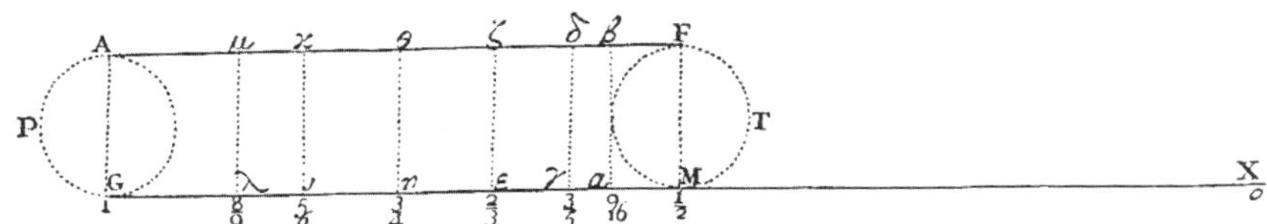

Figure 4   Referred to in paragraph 461.

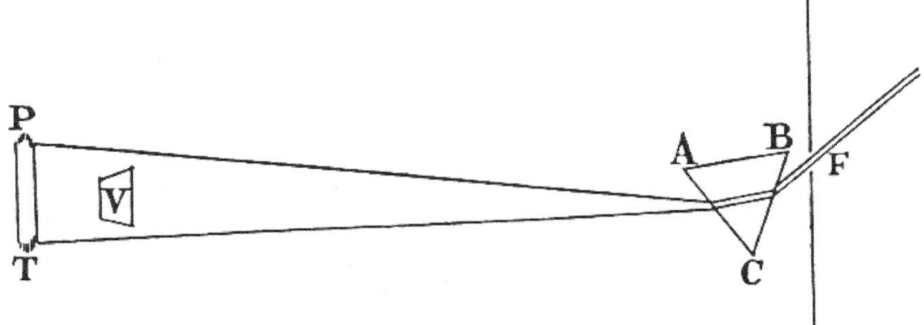

Figure 5   Referred to in paragraph 510.

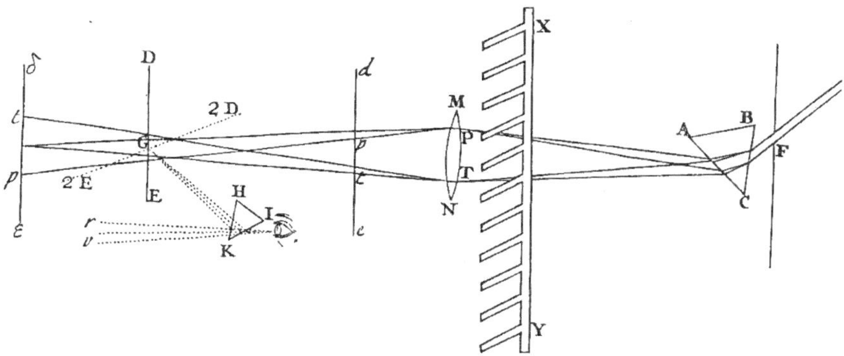

Figure 6   Referred to in paragraph 545.

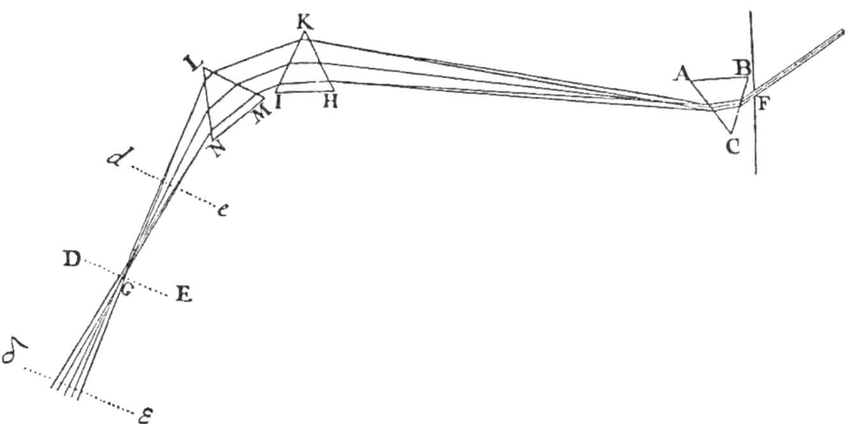

Figure 7   Referred to in paragraph 547.

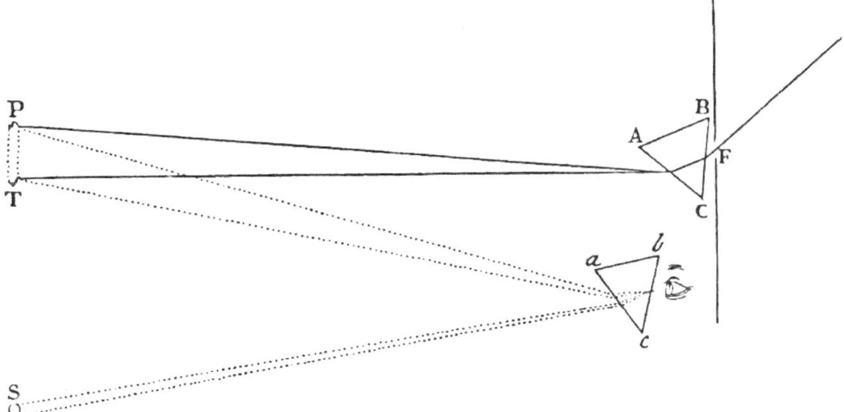

Figure 8   Referred to in paragraph 544.

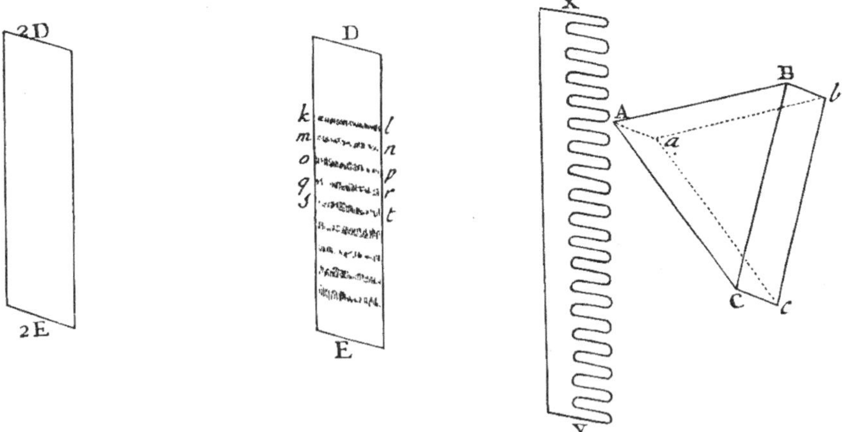

Figure 9   Referred to in paragraph 518.

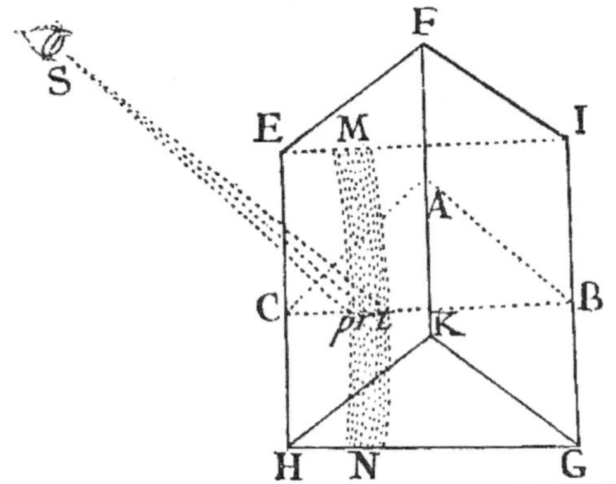

Figure 8   Referred to in paragraph 604.

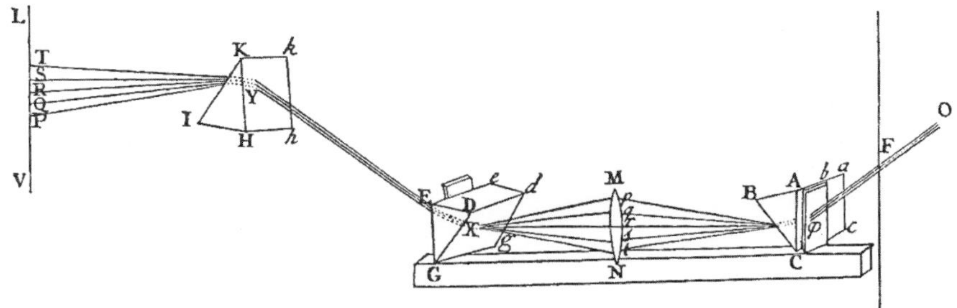

Figure 16   Referred to in paragraph 678.